"We cannot improve beekeeping by going farther and farther away from the bees' natural tendencies. Instead, pick the hive model that is best matched to your locale, populate it with local bees, and the results will speak for themselves."

Georges de Layens

DR. DIANA SAMMATARO
author of *The Beekeeper's Handbook*, USDA Bee Lab scientist emeritus

"What a magnificent and comprehensive work! I truly enjoyed it. Well written and captivating, impressive illustrations and very valuable information. Certainly a complete guide to beekeeping and a glimpse of how simple beekeeping could be if the mites had not changed our bee management practices. I especially like the information on managing fixed-comb hives (too bad they are not legal in the U.S.)—beekeepers' ingenuity never fails to amaze me. The information on bee biology and flowers demonstrates how fascination with bees leads us to investigate their behavior and explore the world around us. Interesting and thorough work, excellent translation. Well worth it, and inspiration to all."

DR. THOMAS D. SEELEY
author of *Honeybee Democracy* and *Following the Wild Bees*

"This is a delightful book full of wonderfully audacious ideas for U.S. beekeepers. Originally written in 1897 by a leading French beekeeper and biologist, Georges de Layens, it presents the design of a movable-frame horizontal hive that was immensely popular in France at the time, and that remains so today in Spain and other parts of Europe. Horizontal hives are often mounted on a stand or equipped with legs, so working bees with these hives feels much like working at a carpenter's bench: easy on the back! Other advantages of these hives include gaining freedom from lifting heavy honey supers; enabling the bees to organize their nest naturally, with brood nest near entrance (at one end) and frames of honey toward rear; and letting the bees live on combs than run continuously from top to bottom, as in nature. I certainly look forward to building and trying out a horizontal movable-frame hive, inspired and guided by reading this fascinating book."

DR. TAMMY HORN
Kentucky State Apiarist
author of *Beeconomy* and *Bees in America: How the Honey Bee Shaped a Nation*

"In the landscape of limited hive diversity, Georges de Layens' book presents a range of time-tested hive models used by European beekeepers to manage their bees. From woven skeps to modern horizontal hives, it covers the diverse hive options and shows how the problem of overwintering can be resolved via hive design. Offering robust beekeeping practices rooted in respect for honeybee biology, *Keeping Bees in Horizontal Hives* is a valuable contribution to the discussion of sustainable beekeeping."

Keeping Bees in Horizontal Hives

A Complete Guide to Apiculture

Georges de Layens
Gaston Bonnier

Translated by Mark Pettus, Ph.D.
Edited by Leo Sharashkin, Ph.D.

DEEP SNOW PRESS
Ithaca • New York

Keeping Bees in Horizontal Hives
A Complete Guide to Apiculture

Georges de Layens
Gaston Bonnier

Translated from the French by Dr. Mark Pettus
Edited by Dr. Leo Sharashkin

Illustrators: A. Millot, P. Jamin, B. Herincq, J. Poinsot, and others
Cover photographs: Dr. Leo Sharashkin
Back cover, 5th photo © Chico Sanchez / Fotobank Lori / age Fotostock

All Rights Reserved
Copyright © 2017 Deep Snow Press

ISBN: 978-0-9842873-6-9
Library of Congress Control Number: 2016963552

Printed on recycled paper with 100% post-consumer recycled content

Published by Deep Snow Press

www.HorizontalHive.com

Contents

A Natural Beekeeping Classic
Editor's Preface *xvii*

PART ONE
INTRODUCTION TO APICULTURE

Preliminary Information
1. Apiculture and its products *3*
2. The future of apiculture *4*
3. Promoting beekeeping *4*
4. The importance of beekeeping for agriculture *5*

Chapter I
Bees
5. Bees at the hive entrance *7*
6. Guards, fanning bees, cleaners *7*
7. Foragers *8*
8. Workers and drones *9*
9. General description of worker bees *10*
10. A description of the drone *13*
11. The young bees' first flight *14*
12. Bees clustering outside the hive *14*
13. Bees on flowers *15*
14. Insects that can be confused with bees *15*

15. How bees visit flowers. Nectar *17*
16. Harvest of nectar other than from flowers. Honeydew *19*
17. How bees harvest pollen *21*
18. Propolis: how bees gather it *23*
19. How bees gather water *23*

Chapter 2
The Colony

20. The bees inside the hive *25*
21. The queen *26*
22. The waxen honeycomb. Cells *27*
23. Worker cells *27*
24. Honey cells. Capped and uncapped honey *27*
25. Pollen cells *29*
26. Cells containing worker bee larvae. The worker brood *29*
27. Drone cells. Drone brood *32*
28. Queen cells *32*
29. How bees build comb *33*
30. New and old comb *34*
31. Division of labor among bees *34*
32. The lifespan of a bee *36*
33. Egg-laying *37*
34. The number of eggs a queen can lay daily *39*
35. A drone-laying queen *41*
36. Development of a worker bee *41*
37. Development of a queen bee *43*
38. Development of a drone *43*
39. Swarming *44*
40. A swarm is cast *44*
41. Primary swarms and afterswarms. The queen's song *47*

Chapter 3
The Hive

42. Fixed-comb hives *49*
43. Suffocating bees *52*
44. Cap hives *52*
45. Stacked hives *54*
46. Movable-frame hives *55*
47. The advantages of frame hives *57*
48. Wax foundation and its advantages *59*

PART TWO
A BEEKEEPER'S APPRENTICESHIP

Chapter 4
A Region's Honey-Producing Potential

49. Assessing local nectar plant resources *65*
50. Wild nectar plants *67*
51. Nectar plants of cultivated fields and meadows *69*
52. Melliferous trees *71*
53. How favorable is a given region for beekeeping? *72*
54. The climate's influence on honey production *73*
55. The soil's influence on honey production *74*

Chapter 5
Setting Up an Apiary

56. The beginning beekeeper and the movable-frame hive *75*
57. Stings. Veil and gloves *76*
58. Pacifying bees *76*
59. Smoking *76*
60. A common smoker *77*
61. Mechanical smoker *77*
62. How to avoid stings. Sting remedies *78*
63. Precautions to prevent your neighbors from being stung *79*
64. What irritates bees *80*
65. Purchasing colonies *80*
66. Buying hives in late summer *81*
67. Assessing the value of hives purchased in late summer *81*
68. Buying hives in late winter *83*
69. When only swarms are available for purchase *84*
70. Prices of hives and swarms *84*
71. Hive location *85*
72. Hive stand. Bottom board *86*
73. The water basin *87*
74. Transporting hives *87*
75. Transporting swarms *89*
76. Wintering purchased fixed-comb hives *90*

Chapter 6
What to Do in the Spring of Year One

77. A beginner's apprenticeship *93*
78. The end of the wintering for hives bought the previous autumn *93*
79. A hive in excellent condition after wintering. Inspecting a fixed-comb hive *94*
80. A weak hive that has wintered well *97*
81. A strong hive that has wintered poorly *97*
82. A hive that has run out of honey *98*
83. A dead hive *98*
84. A hive that is queenless or has a drone-laying queen *99*
85. What should be done with a dead or disorganized hive? *100*
86. Treating combs with sulfur *100*
87. Feeding hives that are low on honey *101*
88. Which hives require feeding? *101*
89. Feeding a fixed-comb hive *102*
90. How bees feed on syrup *104*
91. When should you stop feeding? *105*
92. Robbing *105*
93. How to stop robbing *106*
94. Pollen substitutes *106*

Chapter 7
Installing Swarms in Moveable-Frame Hives

95. The honey flow *108*
96. Various methods for judging how the honey flow is going *108*
97. Preparing movable-frame hives for installing swarms *110*
98. Description of a movable-frame hive *110*
99. Installing foundation *115*
100. Frames primed with comb *118*
101. How to prime frames with comb *118*
102. Priming frames with a bead of wax *119*
103. Final preparations for installing a swarm *121*
104. How to collect a swarm *121*
105. When a swarm is awkwardly located *123*
106. Determining the swarm's hive of origin *124*
107. Installing the swarm in a moveable-frame hive *125*
108. What if you're unable to identify the swarm's hive of origin? *128*
109. Feeding a swarm in the event of bad weather *128*
110. What to do when there is an afterswarm *128*
111. Collecting an afterswarm *129*
112. Determining the afterswarm's hive of origin *129*

113. Returning an afterswarm to its hive of origin *130*
114. Various circumstances you may encounter when swarms emerge *130*
115. The apiary after swarming season *132*

Chapter 8
What to Do during the Summer of Year One

116. Handling an empty moveable-frame hive *134*
117. A frame box *134*
118. Inspecting moveable-frame hives *135*
119. The advantages of foundation frames when a swarm is installed in a moveable-frame hive *140*
120. Monitoring the remaining fixed-comb hives *140*
121. Monitoring moveable-frame hives *141*
122. End of the honey season *141*

Chapter 9
What to Do During the Fall of Year One

123. How a beekeeper harvests honey *143*
124. Inspecting hives at the end of the season and evaluating the weight of honey in a frame *143*
125. How much honey should be left for winter reserves? *144*
126. When the moveable-frame hives are short on honey *146*
127. Feeding moveable-frame hives *146*
128. What if robbing breaks out? *147*
129. Tools required for harvesting honey from moveable-frame hives *148*
130. Extracting honey *151*
131. Hives that are queenless or almost without honey *153*
132. Uniting moveable-frame hives *153*
133. Fall inspection of remaining fixed-comb hives *154*
134. Wintering of moveable-frame hives and fixed-comb hives *155*

Chapter 10
What to Do in the Spring of Year Two

135. End of wintering *158*
136. Inspecting the hives during early spring in year two *158*
137. Judging the quality of the brood *159*
138. What should be done with a disorganized moveable-frame hive? *159*
139. Arranging the frames during the spring inspection in year two *161*
140. Feeding the hives that are low on honey *162*

141. Problems with feeding *163*
142. Transferring fixed-comb hives into moveable-frame hives *163*
143. A flipped-hive transfer *164*
144. Direct transfer *166*
145. Preparing frames to hold comb from the fixed-comb hive *166*
146. Driving bees from a fixed-comb hive you're going to transfer *167*
147. Removing the comb from the fixed-comb hive and placing it in the prepared frames *170*
148. Moving the bees to their new hive *172*
149. What to do with the comb that wasn't used during the transfer *173*
150. What if there aren't any eggs on the black fabric during the transfer? *174*
151. Monitoring the transferred hive *175*
152. Problems with direct transfer. Other transfer methods *175*

Chapter 11
What to Do in the Summer and Fall of Year Two

153. Monitoring hives during honey season *177*
154. Strengthening a weak hive *178*
155. What to do when the combs have collapsed *179*
156. What causes comb collapse *179*
157. Swarm prevention *180*
158. Supersedure (natural queen replacement) *180*
159. Fall inspection, honey harvest, and wintering *180*

Chapter 12
What to Do in Year Three

160. The end of wintering. Year three *182*
161. Arranging your frames in the spring *182*
162. Maintaining or increasing the number of hives *184*
163. Artificial swarming *185*
164. Hives purchased in another region *188*
165. General monitoring of your apiary during year three *188*
166. Tracking your apiary with a table *190*
167. Fall inspection. Harvesting honey and preparing for winter *190*
168. Capped honey reserves *191*
169. Inspecting comb *192*
170. What to do during winter *192*

Summary of Beekeeping Procedures (Simplified Method) *193*

PART THREE
OTHER HIVE SYSTEMS

Chapter 13
Vertical Hive Equipment

171. The vertical movable-frame hive *197*
172. A description of the vertical hive *199*
173. Notes on the vertical hive *201*
174. The advantages of a vertical hive for producing section honey *202*
175. Equipment needed for producing section honey *203*

Chapter 14
Managing Bees in Vertical Hives

176. General observations *205*
177. The end of wintering and the spring inspection *205*
178. Preparing supers *206*
179. When to add the first super *206*
180. Adding the first super *207*
181. Problems that can result from adding the first super too early *208*
182. Problems that can result from adding the first super too late *208*
183. Monitoring your supers *208*
184. Adding the second super *208*
185. More supers *210*
186. Adding supers for the autumn honey flow *210*
187. Inspecting the hives when supers are in place *210*
188. Harvesting the supers *210*
189. Inspecting after harvesting *211*
190. Preparations for wintering *212*
191. Sections. Comb honey *212*
192. How to encourage the bees to fill out the sections with comb *213*
193. Problems to avoid when producing section honey *215*
194. Section comb honey with horizontal hives *215*

Chapter 15
Keeping Bees in Traditional Fixed-Comb Hives

195. General remarks *218*
196. The end of wintering and springtime procedures *219*
197. Swarming season *220*
198. Uniting weak or late swarms *221*

199. How to unite swarms with swarms *221*
200. Artificial swarming with fixed-comb hives *222*
201. Harvesting honey from fixed-comb hives *224*
202. Processing honey *225*
203. Uniting hives after the honey harvest *226*
204. How to unite hives after the honey harvest *227*
205. When fall feeding is required *229*
206. Readying for winter *230*
207. Managing cap hives *230*
208. Frame caps *232*

Chapter 16
Supplementary Material

209. General observations *234*
210. Frame hives similar to those already described *235*
211. Various systems of moveable-frame hive *236*
212. Varieties of horizontal hives *236*
213.1. Varieties of vertical hives *237*
213.2. Hives employing a two-queen system *238*
214. "Warm-way" hives *239*
215. Varieties of fixed-comb hives *240*
216. Choosing a hive *242*
217. Observation hive *242*
218. Advantages and disadvantages of a beehouse *243*
219. Scales, thermometers, hygrometers, barometers, and microscopes *244*
220. Feeders. Feeder varieties *246*
221. A water-bath oil can *246*
222. An uncapping fork *248*
223. Bee repellents *248*
224. Drone traps *248*
225. Varieties of extractors *250*
226. Bee escapes *250*
227. Division boards *251*

Chapter 17
Alternative Procedures

228. General observations *253*
229. Buying populated moveable-frame hives *253*
230. Transfer by superposition or by artificial swarming *254*
231. Speculative feeding *255*
232. Feeding with sugar paste *257*

233. Other methods for preventing afterswarms *257*
234. Artificial swarming with one hive *258*
235. Other procedures for uniting colonies *259*
236. Restoring queenless hives *259*
237. Artificial requeening *260*
238. Requeening by natural swarming *262*
239. Requeening by grafting queen cells. Making nucleus hives *262*
240. Introducing a queen into a hive using a queen cage *264*
241. Introducing a queen into a hive through the entrance *265*
242. Foreign bee races *266*
243. Comb honey without sections *268*
244. Partial harvest from a fixed-comb hive *268*

PART FOUR
GENERAL OBSERVATIONS ON BEEKEEPING

Chapter 18
General Principles and Comparison of Methods

245. Preliminary remarks *273*
246. General principles applicable to all hive systems *273*
247. To what extent should bees be allowed to build comb? *276*
248. Protecting a colony against fluctuations in temperature *276*
249. Various kinds of beekeepers *277*
250. The sideline beekeeper *278*
251. The professional beekeeper *278*
252. The amateur beekeeper *279*

Chapter 19
Apiary Products

253. General considerations *281*
254. The honey room *282*
255. Storing honey *283*
256. Selling honey *284*
257.1 Major varieties of honey *284*
257.2 Honey composition *285*
258. Mead *286*
259. Low-quality mead *287*
260. Alcohol content of a fine mead *287*

261. The bouquet and color of mead *287*
262. General production method *288*
263. The Guyot glucometer *290*
264. Using wash water when making mead *290*
265. Time required for fermentation *291*
266. Fining and bottling mead. Barrels and barrel maintenance *292*
267. The hydrometer *293*
268. Meads with various sugar content *294*
269. Mead composition *294*
270. Ameliorating wine with honey *295*
271. Pyment (grape mead) *296*
272. Pomace wines *296*
273. Cyser *297*
274. Honey vinegar *297*
275. Honey brandy *297*
276. Uses of honey *298*
277. Wax production *298*
278. The solar wax melter *300*
279. Large-scale wax production *300*
280. Making wax foundation *301*
281. Telling artificial wax from real wax *301*
282. Uses of wax *303*

Chapter 20
Bee Diseases and Enemies

283. Foulbrood, or brood rot *305*
284. How to spot foulbrood *306*
285. Hygienic measures for preventing foulbrood *307*
286. Treating foulbrood *308*
287. Disinfecting a hive infected by foulbrood *309*
288. Dysentery *309*
289. Other bee diseases *310*
290. The greater wax moth *311*
291. How bees control the wax moth *312*
292. Other enemies of bees *313*
293. Plants that are harmful to bees *316*

Chapter 21
Nectar and Nectar Glands

294. Nectar glands *318*
295. Sugars contained in nectar glands *318*

296. Nectar contains much more water than honey does *321*
297. Honey has a different composition than nectar *322*
298. Nectar glands outside the flower *323*
299. The nectar glands of the nasturtium, hellebore, and horse chestnut *324*
300. The nectar glands of reseda, violets, peach trees, and legumes *325*
301. The nectar glands of crucifers, anemones, heathers, and buckwheat *327*
302. The nectar glands of periwinkles (*Vinca*), mints (*Lamiaceae*), figworts (*Scrophularia*), and houseleeks (*Sempervivum*) *329*
303. The nectar glands of Scabiosae and Compositae *331*

Chapter 22
The Honey Yield of Various Plants

304. How nectar glands secrete nectar *333*
305. Variation in nectar volume over the course of the day *334*
306. Meteorological variations in nectar yield *336*
307. Variations in nectar yield with respect to soil and air humidity *337*
308. Variations in nectar yield due to soil composition *339*
309. Variation in nectar yield due to climate *339*
310. Honeydew and sugary exudations of plants *340*
311. How bees allocate foragers among honey plants *343*

Chapter 23
Variations in a Bee Colony's Activity Levels Throughout the Year

312. Bee activity over the course of a single foraging season *345*
313. Variations in hive weight over the course of a day *347*
314. The yield is not always proportional to the bees' activity level *349*
315. Honey consumed during winter *349*

Chapter 24
Managing an Outyard *351*

Index *362*

A Natural Beekeeping Classic
Editor's Preface

"Keeping bees requires little effort, and barely any capital to get started," wrote Georges de Layens in *Keeping Bees in Horizontal Hives: A Complete Guide to Apiculture*. Europe's leading beekeeping authority, he certainly knew what he was talking about. Following his methods with my forty hives, I witnessed that keeping bees can indeed be simpler than growing tomatoes, but most beekeepers' experiences today are quite different.

"My mother spent $5,000 on her seven hives in the first few years!" exclaimed Alan, his voice trembling with emotion. A typical scenario: you attend a beginner beekeeper's class, buy equipment and protective gear, order bee packages, install them in the hives, treat against parasites and disease, feed in the fall and then... they do not survive the first winter. You buy more bees the following spring and the cycle repeats itself. Faced with high bee mortality, mounting costs, and modest returns even many experienced beekeepers hang it up. There are half as many bee colonies in the U.S. today as there were in the 1940s.

What gives? Georges de Layens, along with many other authors, documented that beekeeping can be simple, healthful, productive, and profitable. Historical records show that a hundred years ago farmers commonly had sizable apiaries in their backyards. Bees required hardly any care and yet produced a honey crop five seasons out of six. Eva Crane described many beekeeping systems where harvesting honey was the only task a keeper had to perform. Why does it sound like science fiction today, and many experts recommend looking into your hives every two weeks?

Georges de Layens emphasized that sustainable beekeeping rests on two principles: *using local bees* and keeping them in *appropriate hives* that are gentle on the bees and the keeper alike, and require minimal management. De Layens spent two decades developing and perfecting a hive—now bearing his name—and a system that allows you to manage your colonies reliably with two or three hive visits per year.

The Layens hive and beekeeping method became one of the most popular in Europe. There are some 2.6 million hives in all of the U.S. Just one European country, Spain, in an area smaller than Texas, uses over 1 million Layens hives today. And, of course, the core principles of good beekeeping described in this book can be successfully applied to any hive model, including the Langstroth hives you may already have. "You can be a good beekeeper with any hive system," wrote de Layens. "But you cannot be a good beekeeper if you don't know what you are doing."

The Layens hive is a horizontal hive holding 20 large frames (13" long by 16" deep) on one level. The number of frames can be lesser or greater depending on your local honey flows; frames' shape and large size promote good wintering and strong spring buildup. With all the frames on the same level, you have ready access to every part of the hive so you can add or remove frames with minimal disturbance to the bees, and there are no heavy supers to lift, ever.

Horizontal hives are loaded with additional frames during the spring inspection and then opened in the fall for honey harvest. If you think this is not feasible, consider this: Jean Hurpin, a prominent French beekeeper and author, founder of France's largest beekeeping magazine *L'Abeille de France*, modified the Layens hive to enable management with *one* hive visit per year. He writes: "Layens hives are wonderfully suited for remote outyards. In the yards that I visit only once per year, I always find the bees in excellent shape with the hives full of honey, harvesting which becomes my sole task." Even today—despite the new challenges bees are facing—my experience has been similar to Hurpin's: simplified beekeeping really works.

After keeping bees in Layens hives for several years and having read his brilliant book, I felt compelled to make it available in the English language—this book you hold it in your hands. I certainly do not expect everyone to switch over to the Layens hive system, but I do feel that every beekeeper can greatly benefit from this knowledge as it expands our understanding of what is possible in beekeeping, and the range of choices we all have.

Faced with the difficulties of beekeeping today, many give up or stoically continue to pull through. But others go back and seek alternate routes. This book may well be this turnaround for those looking for more sustainable and bee-friendly (and beekeeper-friendly!) alternatives. As many of the conventional approaches ("requeen, treat, feed, start over") either break down or become unreliable, unhealthy, and expensive, I trust that many would love to rediscover a simpler beekeeping that avoids many of the chores and complexities we've been taught to take for granted.

In translating this work of one of the greatest beekeeping minds, we took care to closely follow de Layens' original writing, resisting the temptation to update, edit, or censor it. Thus, for example, we retained all the material on fixed-comb hives—almost an anathema today—and I actually find it enlightening that de Layens clearly saw the traditional skeps and log hives as superior to the modern movable-frame hives in some situations.

I would not follow any advice blindly, either in this book or in any other. For example, as a natural beekeeper I do not try to limit the number of drones in the colony since we now have a better understanding of their importance for the long-term success of the bee species. Nor would I recommend using mothballs (especially naphthalene) to prevent wax moth damage—the same can be achieved by freezing comb, while avoiding contaminating wax and honey with chemicals. A handful of objections aside, however, I do find the core of de Layens' method remarkably modern and perhaps even more relevant for us today than when the book was first written.

For instance, even though de Layens' book was published long before Varroa mites' introduction to Europe and the Americas, his method—relying on local disease-resistant bees, regularly breaking the brood cycle, providing bees with optimal nutrition, maintaining strong colonies, and practicing non-invasive management—certainly represents a much better long-term solution to this pest problem than chemical treatments.

"We cannot improve beekeeping by going farther and farther away from the bees' natural tendencies," wrote de Layens. "Instead, pick the hive model that is best matched to your locale, populate it with local bees, and the results will speak for themselves." I fully agree.

In closing I'd love to express special thanks to David Liedlich from Connecticut, himself a Layens hive beekeeper, first for encouraging me to proceed with this project and then painstakingly reviewing the manuscript and making a slew of useful suggestions for improvement.

Dr. Leo Sharashkin
Ozark Mountains, Missouri

About Dr. Leo Sharashkin

Dr. Leo Sharashkin is editor of *Keeping Bees With a Smile*, a comprehensive book on natural beekeeping, and *Keeping Bees in Horizontal Hives: A Complete Guide to Apiculture* by Georges de Layens. Dr. Leo is a regular contributor to *American Bee Journal*, *Bee Culture*, *The Beekeepers Quarterly* (UK), *Acres U.S.A.* and other major publications, and teaches natural beekeeping at his apiary in southern Missouri and around the country and internationally. His apiaries are entirely composed of survivor stock obtained by catching wild swarms, housed in a variety of horizontal hives, including the Layens. His website (including free hive plans): **www.HorizontalHive.com**

PART ONE
INTRODUCTION TO APICULTURE

Preliminary Information

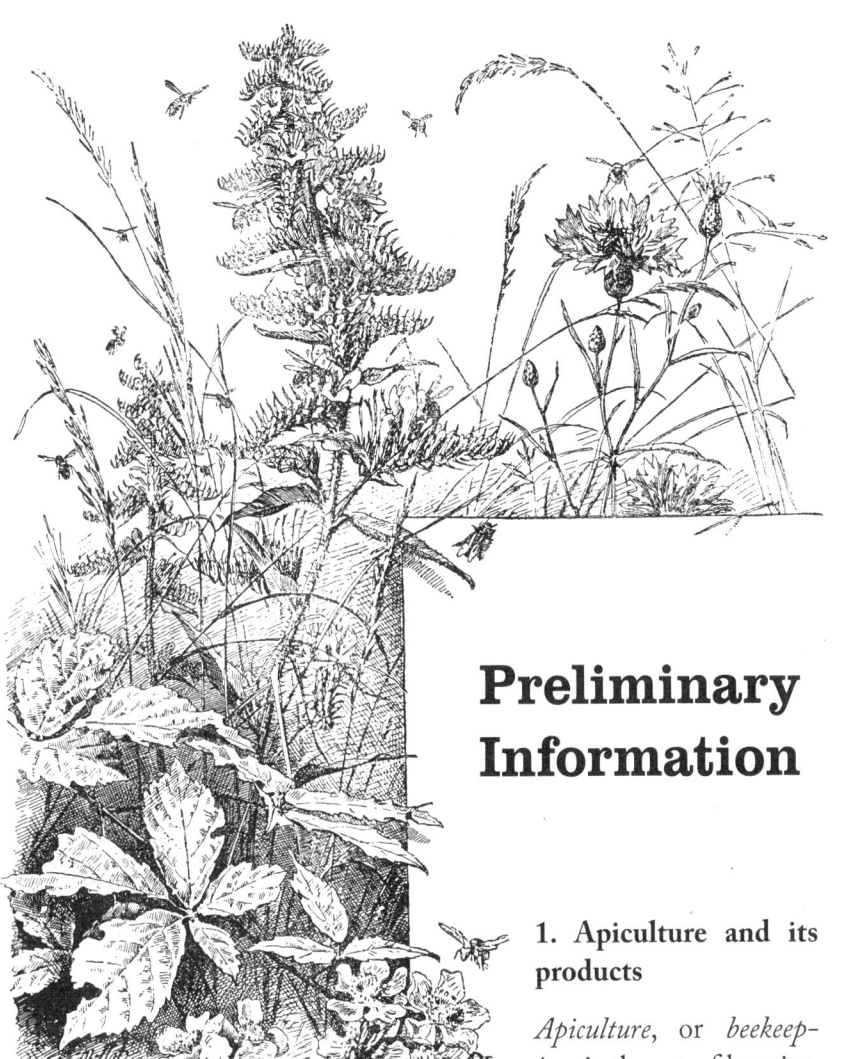

Figure 1. Bees on flowers.

1. Apiculture and its products

Apiculture, or *beekeeping*, is the art of keeping bees in order to harvest *honey* and *wax*.

Honey is by far the more important of these two products. Honey—the sweet substance that bees store in their hives—has multiple uses, and it can be a valuable resource. Not only is honey consumed directly and used as medicine, but it can also serve as a sugar substitute in many cases. Last but not least, by allowing this naturally sweet substance to ferment, we can easily produce *mead*, an alcoholic beverage with all of the health benefits of wine.

Although the potential uses of beeswax are less varied and developed than that of honey, a great quantity of wax is used in industry.

2. The future of apiculture

One shouldn't judge beekeeping's potential by current harvest levels, which average 30 to 35 million pounds (14,000 to 16,000 tonnes) per year.

In France, for example, the number of hives could be increased dramatically, to an extent that is simply impossible to estimate.

In our prairies and heaths, in our fields of sainfoin, buckwheat, and rape, across almost the entire expanse of our mountainous regions and Mediterranean coast, our flowers yield an enormous quantity of nectar—and the vast majority of it goes completely unused.

And for now, at least, people little suspect that this potential source of wealth even exists.

Country folk can tap into this resource without being distracted from their other agricultural pursuits, since beekeeping requires very little work, and only a modest investment to get started.

If French apiculture were to reach its full potential, then beekeepers and, in turn, villagers and factory workers, would have access to a healthy and natural food source, whose origins are easy to verify. On top of that, they'd also enjoy an alcoholic beverage that they could easily produce themselves, avoiding counterfeit and tainted wines.

3. Promoting beekeeping

In many regions of France, many schoolteachers and clergy have already become zealous proponents of apiculture. It is up to them, more than anyone, to encourage a love of bees and to demonstrate beekeeping's potential benefits.

But it's also up to dedicated amateur beekeepers to spread information about the best methods and most appropriate hive models.

We also need to set up a *model apiary* in every teachers' training college and every seminary. Indeed, many such apiaries are already in place. But we must emphasize that such educational apiaries will only be of real service if they are genuine working apiaries.

A permanent collection of every last model of hive, stocked with bees of various foreign races, would be of little real value. Only an apiary that is composed almost entirely of the kind of hive that is best suited to the given region, with local bees, and managed with the express intent of maximizing honey production, will be of real use to potential beekeepers—and they'll be more impressed by the resulting benefits than by any theoretical instruction they might receive. That is, only working apiaries—not ones merely for show—should be the model.

Meanwhile, *beekeeping associations*, ever more numerous in France, are contributing greatly to popularizing beekeeping.

Thanks to all of these devoted efforts, there is no doubt that beekeeping will soon grow to its full potential.

4. The importance of beekeeping for agriculture

Growers' interest in apiculture isn't limited to the important products it yields; it serves them indirectly as well.

Whenever honey plants are grown to produce either grain or fruit, bees will contribute in important ways to boosting the harvest as they fly from flower to flower gathering nectar.

A grower who has hives in his orchard can see his average fruit harvest rise year by year, since, thanks to the bees, the fruit will set in larger numbers. A cultivator who grows rape, lentil, chickpeas, beans, or forage crops will boost his fields' yield if some apiaries are located nearby.

The alleged harm caused by bees feeding on grapes or other sweet fruits is only based on poor observation of the facts. The misconception—unfortunately quite widespread—that bees are harmful in this way cannot be fought strongly enough, since it is a proven fact that bees are incapable of puncturing the skin of these fruits on their own; they merely suck the juice from these fruits once they have already been broken open by birds, wasps or hornets.

In sum, bees do no harm to growers—on the contrary, they often do a great service by boosting the yield of many crops.

Figure 2. Bees at a hive entrance. 1. A guard, recognizing a worker who is returning to the hive. 2. Fanning bees. 3. A cleaner pulling a dead bee out of the hive. 4. A forager loaded down with honey, resting on a plant before re-entering the hive. 5. A forager returning with pollen. 6. Drones.

Chapter 1
Bees

5. Bees at the hive entrance

To get an initial idea of what bees do, let's observe them at the entrance of a strong hive on a nice June morning, when they're highly active. If we stand unobtrusively near the hive, avoid any abrupt movement and keep quiet, we need not worry about being stung.*

6. Guards, fanning bees, cleaners

Guards. Let's first direct our attention to the bees at the hive entrance; we can see several of them pacing back and forth in front of the door, apparently very attentive to anything approaching from outside. We see them checking the returning bees—even trying, it would seem, to recognize them as they arrive (1, fig. 2); they only seem to admit the foraging bees after a kind of inspection. Indeed, we'll often notice that certain bees who look very much like the others, but seem timid upon arriving at the entrance, are chased and driven away by the guards. These uninvited guests actually belong to other hives, and would like nothing more than to sneak into the hive we're watching in order to steal its honey. This scene is even more dramatic when a wasp, hornet or bumblebee attempts to infiltrate the hive.

Sometimes, in certain regions, bees are forced to defend themselves against an even more dangerous enemy, the death's-head hawkmoth, which also tries to steal honey from beehives (see § 292).

---

* If you fear being stung, you can wear a veil and gloves (see § 58).

The bees who police the hive entrance are called *guards*.

Fanning bees. At this time of year, especially in the evenings after a good day's harvest, we can also see other bees alongside the guards, but who, unlike the guards, remain in one place, with their heads turned toward the door, propped up high on their legs, often lined up behind one another, beating their wings at such a high rate that one can hardly see them (2, fig. 2).

As we can easily see, it's this rapid agitation of these bees' wings that causes the peculiar humming sound heard near highly active hives late in the evening.

These bees pay no mind to the comings and goings of the foragers, focused completely on their own special function: to maintain a powerful current of air in the hive's interior by beating their wings.

The greater the harvest has been during a given day, the more of these bees appear come evening.

They are called *fanning bees* or *ventilating bees*, since their role is to ventilate the hive.

Cleaners. If we arrive in the morning, when the bees are just beginning to get to work, we can find yet another kind of bee in front of the hive entrance, busy removing useless debris out of the hive, or discarding at a distance the corpses of any bees who have died during the night (3, fig. 2).

Speaking broadly, one might refer to these bees as *cleaners*.

7. Foragers

Now let's consider the bees regularly passing in and out of the hive as they go foraging.

We're struck, first and foremost, by the feverish activity of these industrious insects as they do their work.

As soon as they emerge from the entrance, these bees head out, without hesitation, in a specific direction. Based on the previous days' harvests, they know in advance where to go to find what they need.

Next, let's take a look at the returning bees: if the flowers are yielding a lot of nectar, then most of the bees will be seen landing somewhat heavily, as if exhausted, on the bottom board supporting the hive, or even on the grass in front of it (4, fig. 2)—these bees are loaded down with the sugary liquid they've harvested from the flowers in order to make honey.

We can also see other bees returning home carrying two small bundles—usually orange or yellow, sometimes pink, white, or various other colors—attached to their back legs (5, fig. 2). These balls are made of pollen (see fig. 8 and § 17) that the bees have gathered from flower stamens and stuck to their legs to carry back to the hive. The pollen is used in the hive as food for the young still developing bees.

Speaking generally, once again, all of the bees who leave the hive for harvesting are referred to as *foragers*.

8. Workers and drones

All of the bees we've seen so far—guards, fanning bees, cleaners, and foragers—are similar in appearance, and are referred to by the blanket term *workers*, or neuter bees (fig. 3).

During the season we've chosen to observe our hive, and especially during the afternoon, we may notice certain bees that are much larger than the others (6, fig. 2); these large bees have a different role to play. They seem to have left the hive for no other reason than to go for a spin; when they return, they don't land ponderously on the hive's bottom board, and make their way into the hive at a leisurely pace. We never see these corpulent bees bringing back any pollen. Indeed,

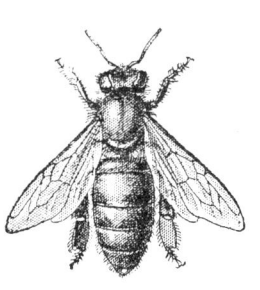
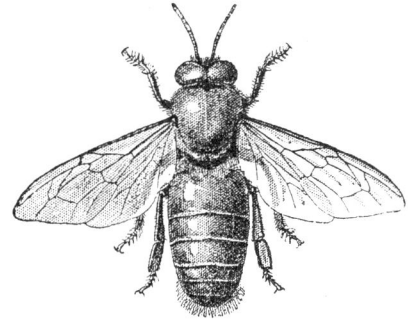

Figure 3. A worker bee (1/2 larger than actual size).

Figure 4. A drone (1/2 larger than actual size).

they do no work at all, and don't even visit flowers. These are *drones*, or male bees (fig. 4). They are rarely seen except during spring and summer.

9. General description of worker bees

We can easily spot some dead bees around the hive; let's pick one up and have a look. We can see that the bee's body, like that of all insects, is divided into three major segments—the *head* (*h*, fig. 5), the *thorax* (*th*), and the *abdomen* (*a*)—and has six legs, all attached to the bottom of the thorax. The bee also has four transparent wings, the two front ones larger than the others, and all crisscrossed by several veins. The wings are attached to the top of the thorax (figs. 3 and 5).

If we examine the head closely (fig. 6), we notice two thin extensions on top, subdivided into smaller segments—these are the bee's *antennae* (*an*, figs. 5 and 6), which serve as olfactory organs.

To the right and left, we see two large, round structures located on the sides of the head—these are the *eyes* (*E*, fig. 6), whose surface a magnifying glass would show to consist of a great number of small, regularly-shaped facets. Between the two eyes, on the top of the head, one can see—again under a magnifying glass—three other eyes, extremely small and smooth, arranged like the three points of a triangle (*e*, fig. 6). It is believed that these three tiny eyes primarily aid the bee in seeing nearby objects.*

Figure 5. A worker bee on a blueweed flower (3 times magnification): *h*, head; *th*, thorax; *a*, abdomen; *an*, one of the two antennae.

At the bottom of the head is the mouth. We notice two main parts: first, two powerful jaws that move from right to left, called *mandibles* (*m*, fig. 6); and, second, an elongated mouthpart that can be retracted into a sort of sheath. This part is called the *proboscis* (*p*, fig. 6)—that is, the bee's tongue.

The bee uses her mandibles to knead wax, to open flower stamens in order to gather pollen, to pick up debris for removal from the hive, or to seize strange insects that attempt to enter her home. The proboscis is used to draw the sugary liquid that will eventually be turned into honey, as well as the water that bees gather.

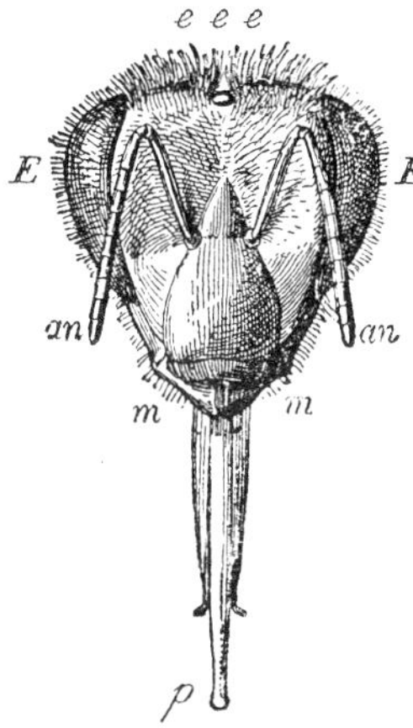

The thorax, as we have seen, supports four wings on top and six legs on the bottom. The two rear legs feature almost spoon-like depressions; it is in these two small hollows, called *pollen baskets* (*B*, fig. 9), that the bees place their balls of pollen (*pn*, fig. 8), using their two front pairs of legs. It's easy to see that the legs are covered with rows of bristles called *combs* or *brushes* (*b*, fig. 7), which are also useful for harvesting pollen.

One also notices that the ball of pollen (*pn*, fig. 8) is held in the basket by curved hairs (*B*, fig. 8) on the side of the basket (see also fig. 9).

Figure 6. A head of a worker, viewed from the front, showing the large eyes, *E*; the small eyes, *e*; the proboscis, *p*; the mandibles, *m*; and the antennae, *an* (7 times magnification).

Now, let's examine the abdomen (fig. 10). We can easily see that it is composed of six rings that, though rigid, can slide slightly atop each other. On their lower section, and beneath them, one can sometimes see a kind of grease emerge, which hardens to form small, very thin scales (*s*, fig. 10). This is the *wax* the bees use to build honeycomb.

---

* The exact function of the three simple eyes, or *ocelli*, remains unknown. They cannot focus or capture images, but are sensitive to light intensity. They may play a role in the bee's circadian rhythms, orientation, or flight control. *Ed.*

The wax is produced by a large number of small glands whose openings (such as p, fig. 11) are located above plate-like formations called *wax mirrors* arranged in pairs in the slits between the abdomen rings. Figure 11 shows a pair of these wax mirrors.

Finally, at the very tip of the abdomen we find the *sting*, which the bee uses to pierce her enemy before injecting venom. Figures 12 and 12.2 show the tip of the abdomen, bisected lengthwise. In figure 12, the bee is not using her sting, s, which is retracted into the sheath, sf. In figure 12.2, the sting is extended; here, one can see how the sheath, sh, is drawn aside to expose the sting. Now free, the sting springs outward, through the gap between the two rearmost plates of the abdomen, and secretes a drop of venom, v, from its tip. This venom is composed of a mixture of liquid from the reservoir, vs, and the glands, $g.ac.$, and of liquid produced by the glands, $g.al.$ Taken separately, each liquid—one acidic, the other alkaline—is harmless, but they become venomous when mixed. Small barbs in the end of the sting cause it to catch in the wound. If the bee is forced to pull free quickly, the sting is ripped off, tearing the abdominal organs and killing the bee. But if the bee is unthreatened, she will have time to withdraw her sting safely.

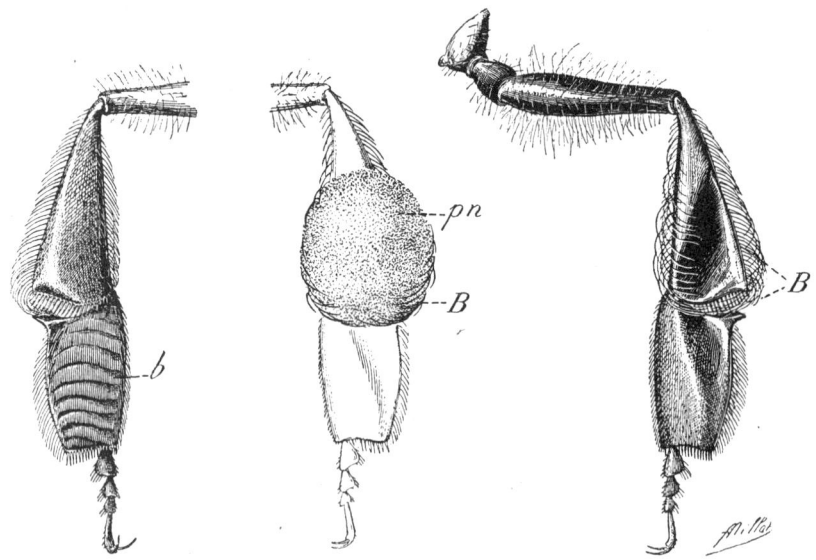

Figures 7, 8, 9. Back legs of worker bees (10 times magnification). Figure 7 (left). Leg, inner side, b, brush. Figure 8 (center). Leg with pollen ball, pn; B, basket hairs retaining the pollen. Figure 9 (right). Leg, outer side; B, basket.

CHAPTER I. BEES 13

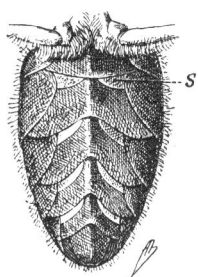

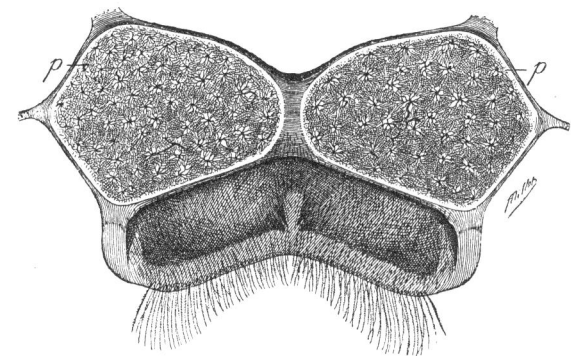

Figure 10. Worker abdomen, viewed from below: *s*, a scale of wax.

Figure 11. Two wax mirrors of an abdominal ring (16 times magnification): *p*, two wax mirror openings.

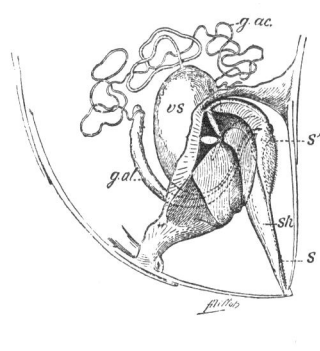

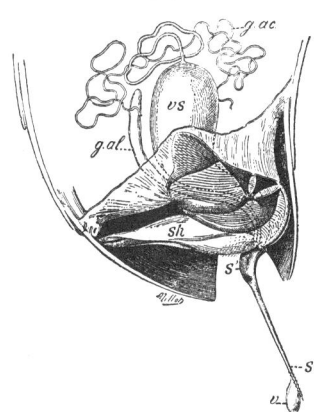

Figure 12. Worker bee sting in its sheath: *s, s'*, sting returned to its sheath, *sh*; *g.ac.* and *g.al.*, glands secreting acidic and alkaline liquids, which, when combined, form the venom; *vs*, venom sac or acidic gland reservoir (10 times magnification).

Figure 12, part 2. Worker bee sting extended from its sheath: the sting, *s, s'*, is extended through the opening between the two rear plates of the abdomen, with the sheath moved upward; *v*, a drop of venom.

10. A description of the drone

If we find the body of a drone near the hive, then a similar examination shows that it differs from the worker bee not only in terms of its

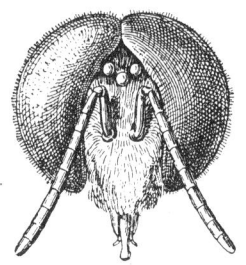

Figure 13. Head of a drone (7 times magnification).

size, but also its shape (see fig. 4). A drone is hairier, lacks the spoon-like indentation in his rear legs, and has neither a sting nor wax glands; his wings are larger and produce a different kind of buzzing sound when they beat. In addition, his head (fig. 13) has a different shape. The two eyes on the side of its head are larger and meet in the middle; the three eyes on the top of the head are positioned further forward.

11. The young bees' first flight

If it's a warm day, we can often see a fairly large number of bees near the hive, flying in a peculiar manner. They don't quickly head off far from the hive like the bees leaving for harvest, but remain circling in front of the hive at various distances, with their heads generally facing the hive entrance.

These are the newly emerged bees that are leaving the hive for the first time and are learning to recognize their home.

The bees engaged in this exercise are said to be performing the *play flight*.

The play flight can take place at various hours of the day, whenever it's hot.

This flight of young bees occurs for almost all hives during the hot spells of summer and accompanies the first flights of spring. It can also follow several consecutive days of rain.

One shouldn't confuse the young bees performing the play flight with the coming and going of the busy foragers. One may, for example, notice bustling activity in front of a hive that is due solely to the flight of young bees, without any workers engaged in harvesting nectar.

12. Bees clustering outside the hive

During spells of intense heat, one often notices a multitude of bees spilling out of the hive entrance; the worker bees hang onto one another with their legs, dangling in groups from the entryway and even beneath the hive. In layman's terms, this is called *bearding*.

Bees form a beard when they run out of room to distance themselves from one another inside the hive when it's extremely hot.

With their home no longer able to accommodate their expanding volume, the hive's residents are forced to spread outside.

13. Bees on flowers

Now let's head into the fields, or into a clearing in the forest, to see how the bees go about their harvest (fig. 1). How do they draw their sugary liquid, and how do they collect their pollen?

Getting close to the bees will be even easier now than it was near their hive, since *bees never sting far from their home*. To force them to deploy their sting while on a harvesting run, one would have not only to grab them, but also squeeze them.

14. Insects that can be confused with bees

First things first—let's not confuse honey bees with other nectar-collecting insects that may resemble them to varying degrees.

On flowers, one often sees a kind of fly of the same color as bees, and with a body only moderately larger. This is the *European hoverfly*, or "drone fly" (*Eristalis tenax*, fig. 15), which one can easily recognize by the fact that it has only two wings, and, consequently, a different

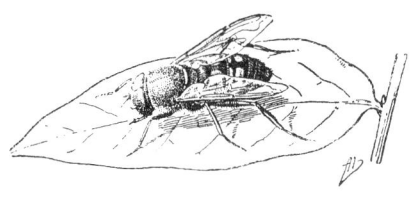

Figure 14. An *Osmia* (*Osmia fronticornis*) on an apricot flower (1/3 larger than actual size).

Figure 15. A drone fly (*Eristalis tenax*), with coloration similar to that of a bee, which is often mistaken for a bee (1/3 larger than actual size).

manner of flight: the buzzing sound it makes differs from a bee's. We can also distinguish it by its shorter antennae and the absence of pollen baskets on its legs.

Other insects that may be mistaken for bees are, like bees, nectar-gathering members of the *Hymenoptera* order. Like the bee, these insects have four transparent wings with rather thick veins, and a tube-shaped tongue for sucking up the sugary liquid of flowers.

The genus *Osmia* includes a certain number of species that may be confused with bees—for example, the *Osmia* shown in figure 14. We can tell them apart from a worker bee by the following traits: they lack baskets on their legs, and gather pollen using a brush located on their ventral side, and their abdomen is angular on the back, not flat like that of the bee.

Figure 16. An *Anthophora* (*Anthophora pilipes*) atop its nest (1/3 larger than actual size).

Figure 17. An *Eucera* (*Eucera longicornis*) gathering nectar from a *Lotus* flower (1/3 larger than actual size).

The genus *Anthophora* (fig. 16) is composed of nectar-gathering insects who build their nest on the ground or in walls; such nests consist of a curved tube fashioned by these insects. They can be distinguished from bees by their extremely hairy body, by the higher pitch of the buzzing sound they make, and especially by the way they visit flowers: *Anthophora* lands on a flower very lightly, then passes to the next one with a liveliness all its own. In the time it takes a bee to visit just a flower or two of a given plant, an *Anthophora* will visit between ten and fourteen.

The genus *Eucera* is quite similar to *Anthophora*; the males are easy to spot by the great length of their antennae (fig. 17).

The genus *Megachile* (fig. 18) includes a large number of species, several of which can also be confused with bees. Like *Osmia*, they have a brush on their underside, but their abdomen is more or less flat on top, and features rings that glide very loosely atop one another. These are the *Megachile* that cut out neat, circular bits of leaves that they use to construct the walls of their nests.

The genus *Chalicodoma* (fig. 19) includes close relatives of the *Megachile*; they are commonly referred to as "mason bees," since they build their nests in walls or rocks.

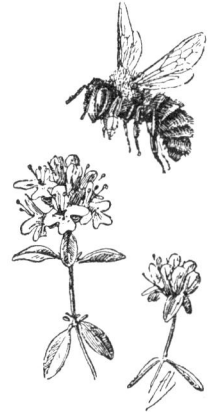

Figure 18. A *Megachile* (*Megachile circumcincta*) approaching wild thyme flowers (actual size).

Figure 19. A *Chalicodoma* (*Chalicodoma rufitarsis*) gathering nectar from a peach blossom (actual size).

15. How bees visit flowers. Nectar

On a nice June day, we will now be able to easily recognize, without mistake, worker bees foraging on the flowers.

Standing before a field of blossoming sainfoin, we notice a large number of bees on the pink clusters of flowers; let's single one out and watch her closely (fig. 20). She alights on one of the flowers in the cluster, spreads aside the petals of the corolla, and sticks her head inside the flower, proboscis extended; then she moves on to the next flower in the cluster and repeats the same procedure. When she reaches a still unopened bud, she flies away immediately to a new cluster, which she works over flower by flower, and so on and so forth.

If we pick a sainfoin flower that has yet to be visited by a bee, delicately pulling the petals aside, we notice a small drop of glistening liquid at the bottom of the flower; we need only put this drop on our tongue to confirm its sweetness. This is the sugary liquid called *nectar* that the worker bees gather and use to make honey.*

Figure 20. A bee working a sainfoin flower.

Figure 21. A bee foraging on a white clover flower.

On cabbage, rape, and turnip blossoms we can see, without even touching the flower, brilliant droplets of nectar gathered there, which a bee can harvest very easily from these plants. On white clover (fig. 21), we can watch the foragers as they extend their proboscis into the interior of the flower's delicate corolla—and if the nectar flow is heavy, bees will even come to collect the nectar overflowing between the corolla and the calyx of the flower.

* For more details, see § 294 and on.

Near a field of beans we can observe a curious phenomenon: since bees' proboscis is often too short to reach the inside of the flower directly, we may see them drawing the nectar through a hole or two cut through the calyx or corolla. It's not the bees who cut these holes; their mandibles, too weak to pierce even the skins of fruit, aren't strong enough to chew through the calyx or the petals. By observing the surrounding plants, we notice that it's the wild bumblebees, like the *large earth bumblebee* (*Bombus terrestris*) or the *common carder bee* (*Bombus pascuorum*), who chew through these flowers in order to harvest the nectar (fig. 22); in so doing, they inadvertently help their competitors, the honey bees, who make use of the holes they've cut.

If we extend these observations to include numerous plants visited by the foragers, we notice that the bees' distribution of labor is highly organized; we can only admire how methodically they spread out among the flowers, in numbers proportional to the harvest to be found there.

We can note another interesting fact: a single forager will only visit a single type of flower during her trip outside the hive.

Figure 22. A bumblebee piercing a bean blossom, allowing bees to visit the flower later.

16. Harvest of nectar other than from flowers. Honeydew

Bees don't only collect sugary liquids from flowers.

In the spring, in a field of vetch, we might be surprised to see large numbers of worker bees busy with their harvest, even though not a single flower has bloomed there yet.

Let's walk closer: we see that the foragers are using their proboscis to gather the numerous droplets of nectar that form at the base of this plant's leaves, in the small hollows located on those peculiar formations which botanists refer to as stipules (fig. 23).*

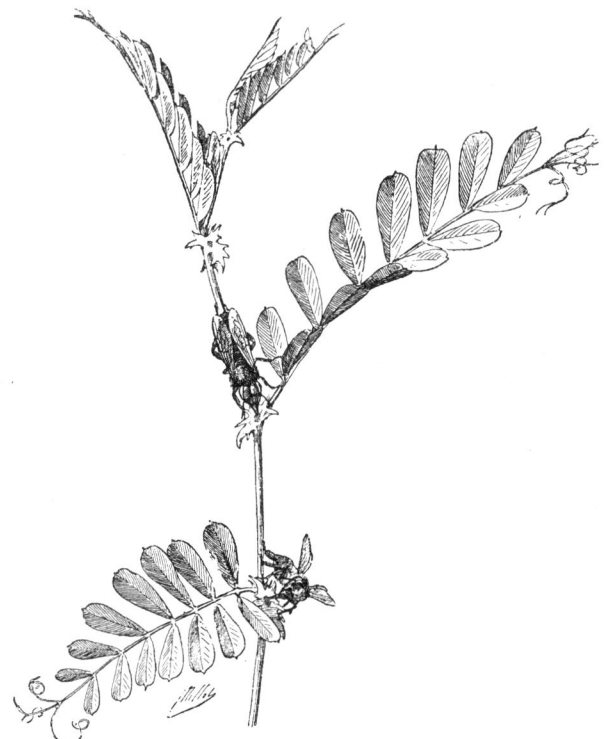

Figure 23. Bees harvesting nectar from vetch stipules.

Figure 24. Bees gathering honeydew from oak leaves.

This nectar is more abundant than that of many flowers, and it proves extremely useful to bees in early spring.

In summer, in the forest, we can also often hear an intense buzzing sound that reaches as far as the upper branches of oaks, birches, beeches, poplars, lindens, firs, and many other trees.

Approaching the lower branches, we can see numerous bees harvesting a sugary liquid on the surface of the leaves (fig. 24): this is *honeydew*—an important resource for bees during hot weather in many forested regions.**

17. How bees harvest pollen

We've seen that bees also harvest pollen. Let's examine how they collect it.

Generally speaking, a given worker bee doesn't harvest nectar and pollen simultaneously. If we watch various flowers, here and there, we see some bees that, instead of sticking their head into the corolla, move actively about the flower's surface, where the *stamens* are located.

As we know, a stamen (fig. 25) consists of a small filament (*f*) crowned by a larger *anther* that contains a colored powder called *pollen* (*p*), which usually exits the anther through two slits. Pollen is required for the formation of seeds.

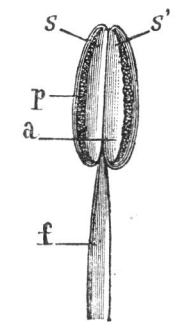

Figure 25. A stamen: *f*, filament; *a*, anther; *s*, *s'*, the two pollen sacks of the anther; *p*, pollen.

Let's examine one of the bees we find on the stamens of an apple tree blossom (fig. 26); we see her gathering the pollen with her mandibles and, when necessary, forcing open the anthers. By kneading the pollen, she produces a small ball that, using her front legs, she then passes backwards, to her right and left, reaching all the way to the baskets on her rear legs.

---

* For more details, see § 298.
** For more details, see § 310.

If the flower has many open stamens, and its entire surface is covered with pollen powder, one can see the worker bee gathering the pollen not with her mandibles, but with the brushes on her legs (see figs. 7, 8, and 9); the bee takes the pollen with the brush on her rear right leg in order to place it in the basket on the rear left leg, and vice versa.

When a lot of pollen has accumulated in one spot, or when the stamen filaments unwind explosively, as on the broom plant, for example,

Figure 26. Bees gathering pollen from apple blossoms.

the bee's entire body may be covered with pollen. When the broom plant is blooming, one can often see bees that have a different color from other bees, because their entire body is yellow with pollen.

As was the case with nectar, a single bee will generally only harvest a single kind of pollen during a given trip.

One shouldn't get the idea that bees harm plants by making away with such large quantities of pollen. Quite the opposite: by visiting flowers, bees often transfer the pollen powder from the stamen to the *stigma*, a small, sticky surface above the flower's ovary, in which seeds develop. Seeds cannot be produced unless some pollen makes it onto the stigma. It is the bee's transfer of pollen to the stigma that makes these insects so beneficial to agriculture. And since the stamens always produce a lot more pollen than is necessary for the plants to be fertilized, the amount of pollen taken to the hive is but an insignificant share of the pollen flowers produce.

18. Propolis: how bees gather it

There's another substance we have yet to speak of that bees bring back to their hive less often. If we take a moment to watch the hive entrance, we may notice that from time to time a worker bee returns home with her baskets filled with two small balls of a resinous substance, translucent and extremely sticky. This isn't pollen, but something called *propolis*. Bees use this substance as a kind of putty for

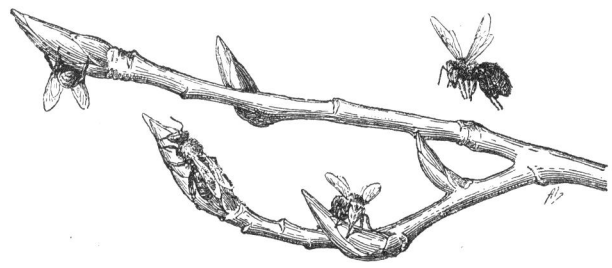

Figure 27. Bees gathering propolis from poplar buds.

strengthening honeycomb or plugging cracks, or even to create a kind of varnish to coat the inside of their hive. Bees gather the resins and gummy substances that make up propolis from the buds of various trees, not from flowers.

Primarily, one can see bees gathering this substance—and sometimes even separating the scales of the buds that are coated with it—from poplars (fig. 27), alders, birches, willows, elms, pines, firs, and other trees.

19. How bees gather water

Bees also bring water back to their hive; they use it to thin down the food they give to the developing young bees, or to dissolve crystallized honey. One can often see them in the morning ingesting drops of dew with their proboscis, or gathering water on the edge of puddles or streams (fig. 28). Since water is indispensable for bees, keepers in regions where surface water is lacking must place a water basin for the bees to use.

Figure 28. Bees gathering water at the edge of a puddle.

Summary

Workers and drones
At the hive entrance one can see bees engaged in various tasks. Some—the guards—keep watch at the hive entrance; others—the fanning bees—are often stationed in rows, beating their wings to produce a current of air in the hive; still others—the cleaners—carry useless debris from the hive. Finally, there are the foragers (the most numerous), which, on a nice day, are busily flying from or back to the hive. All of these bees are called *workers*.

At the hive entrance, you may also notice larger and less active bees: these are the *drones*, or male bees. Drones are distinguished from workers above all by their larger bodies, by the different buzzing sound they make, and by their rear legs, which lack the spoon-like depressions seen on those of worker bees. Drones also lack a sting.

Materials gathered by bees
When forager bees leave for their harvest, they gather:
1. *Nectar*, which they use to make honey.
2. *Pollen*, which is used to feed the young bees.
3. *Propolis*, which is used as a putty to plug cracks or strengthen honeycomb.
4. *Water*, which is used to thin out the food given to young bees, or to dissolve crystallized honey.

Chapter 2
The Colony

20. The bees inside the hive

We've examined the bees near the hive entrance, and we've seen how the foragers gather nectar, pollen, propolis and water. Now, it's time we tried to grasp how things work inside a bee colony.

We need to understand the various jobs these insects perform inside their hive, how they build their home, store their provisions, lay their eggs, and raise their young, which in turn maintain or expand the colony's population.

Observing the bees at work inside their home requires some ability in working with bees and their hives. For now, let's assume we already have the necessary experience, and turn over a traditional fixed-comb hive, like those usually found in the countryside (fig. 58), having properly smoked it first (fig. 97), and taking all precautions to avoid being stung (see § 57). Or, we could open a moveable-frame hive (§ 67)—that is, a hive from which we can remove and examine, one by one, frames containing the waxen honeycombs that the bees have constructed. Or, finally, we could resort to another research tool that requires no special preparation: an *observation hive*. The best observation hive is a small hive with just one layer of honeycomb placed between two panes of glass, covered with shutters that can be opened (fig. 29).

Thanks to these tools, we can easily see, first, that the hive's interior consists of large sheets of wax (fig. 32) full of regularly spaced cavities (fig. 33). These sheets are called *combs*, and each of the small, regular cavities is called a *cell*. Between any two honeycombs inside a hive, a

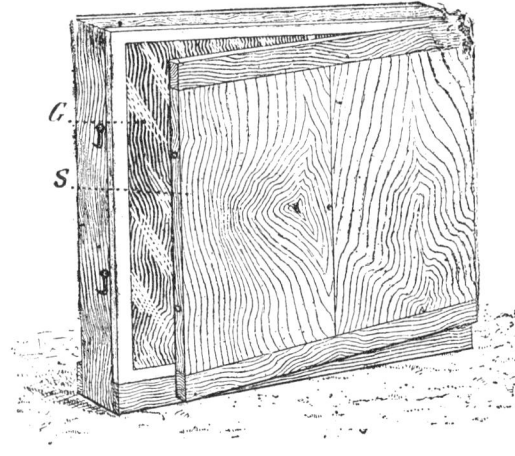

gap of about 3/8" (9–10 mm) is left.

It is in this space between the combs that one sees the bees crowded rather tightly against one another, busy with their various tasks.

Figure 29. An observation hive: S, shutter; G, glass pane.

21. The queen

But before we describe the hive's inner organization, we must understand that the colony is home to one very special bee that we have yet to mention. She lays all of the colony's eggs herself. We refer to this bee as the *mother bee*, or by the more common, but poorly justified, title of *queen bee*.

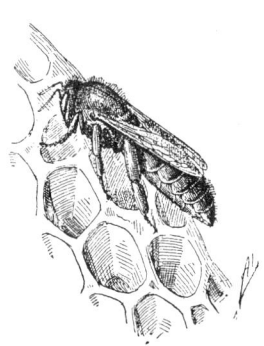

An experienced beekeeper is able to locate the queen bee in any given hive. If we look through the glass of an observation hive, we can see her, surrounded by a retinue of bees.

The queen (figs. 30 and 31) is larger and, more noticeably, longer than a worker bee, and her wings are relatively shorter. Her body has a stronger reddish tint and is shinier on top, and more yellowish on the bottom. When a queen is quite advanced in age, she may be almost blackish in appearance.

Figure 30. A queen bee on a comb (1/3 larger than actual size).

Since her only function is laying eggs, one can understand why her legs aren't built for harvesting like those of the worker bees. In fact, we don't see any brushes or baskets, not to mention the wax glands on the bottom of her abdomen. Her stinger is more curved than that of a worker bee, and she only uses it on very rare occasions. One can take a queen bee in hand without being stung.

22. The waxen honeycomb. Cells

Let's begin by inspecting the shape of the waxen structures inside the hive that we refer to as combs.

If we take a look at an ordinary comb—especially on the side of the hive—we can see some cells (*D*, fig. 33) that are larger than the others. These larger cells are used to raise male bees, or drones; the smaller cells are used to raise workers (*W*, fig. 33).

Figure 31. Queen bee (1/3 larger than actual size).

23. Worker cells

Now, let's examine the worker cells in detail. They generally consist of six equal sides, and are therefore shaped like a six-sided prism whose far end is made of three oblique faces.

If we cut the comb perpendicularly, we'll see that these prisms are built at a slight angle (fig. 34), such that their entrance is positioned somewhat higher than their bottom. This prevents the honey from spilling out; moreover, the axis of a cell on one side of the comb lines up exactly with the junction of three cells on the other side. The walls of these cells are made with *wax*, a substance that, as we have seen, is produced by special glands located on worker bees' abdomens.

24. Honey cells. Capped and uncapped honey

At first glance, we notice that all worker cells, identical in terms of shape, can contain various products.

At the very top of the honeycomb, and moving down the sides to both right and left, we see cells that are sealed with an extremely thin cap, often slightly depressed, as if one had pressed down on it with one's fingertip (*c*, fig. 35).

Let's remove this cap with our fingernail. Underneath, we find a

COMB & CELLS. BROOD COMB

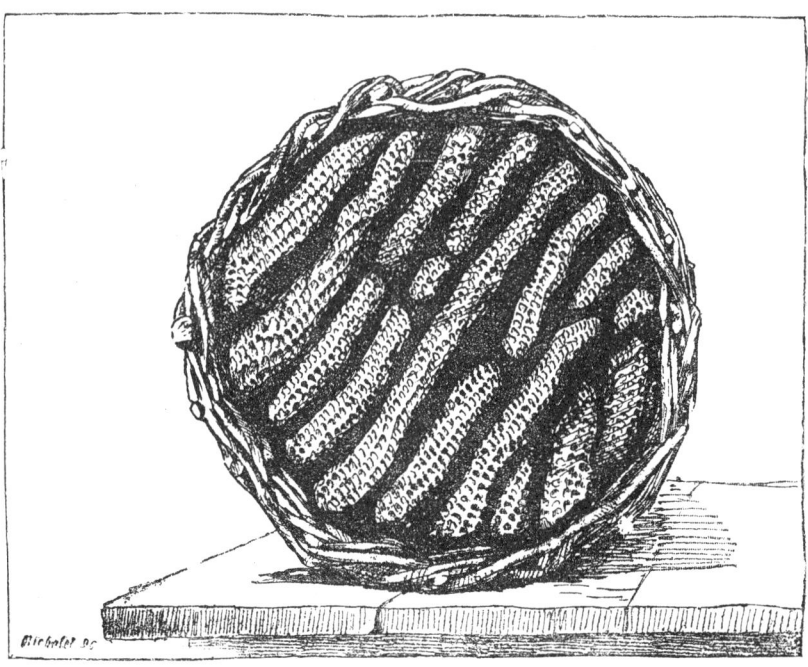

Figure 32. Skep viewed from the bottom, with the ends of the combs visible.

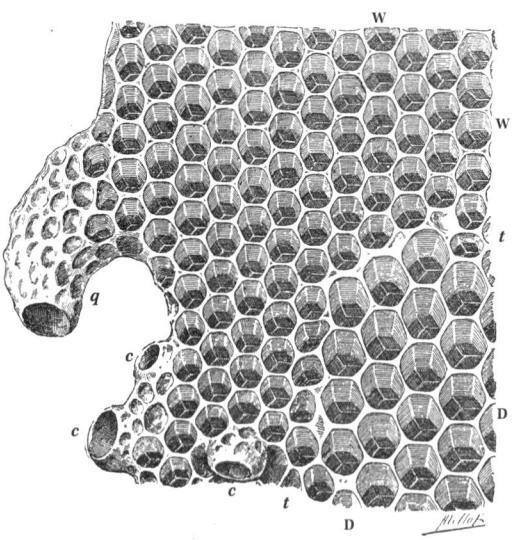

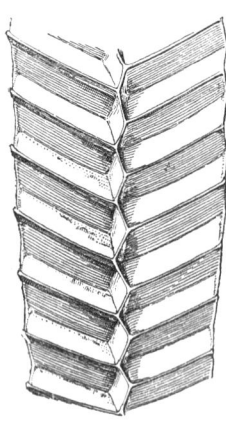

Figure 33. Piece of honeycomb, showing worker cells, *W*; drone cells, *D*; transitional cells, *t*; a completed queen cell, *q*; and some queen cells still under construction, *c* (actual size).

Figure 34. Lengthwise cross-section of comb showing the angle of the cells (actual size).

cell filled with a thick and fragrant liquid; we taste it—it's very sweet. This is *honey* that the bees have stored in these cells and sealed off with a cap referred to as *capping*. Before sealing off a honey cell, worker bees have been observed to add a tiny drop of venom from their sting; this venom contains formic acid that prevents the honey from deteriorating.* But not all of the cells that contain honey are filled to the top, and not all are capped; indeed, a bit further down, we notice cells that are open and only partially filled with honey (*h*, fig. 35)—and the less full a given cell is, the more water its honey contains. The bees will only cap a cell when it's sufficiently full and has reached the desired sugar concentration.

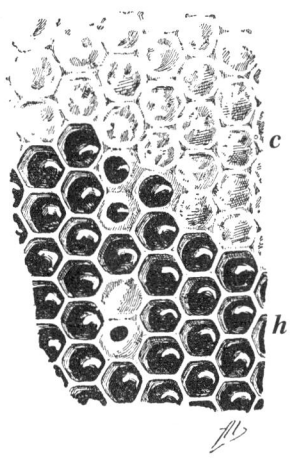

Figure 35. Piece of comb with cells containing capped honey, *c*, and uncapped honey, *h* (actual size).

25. Pollen cells

Here and there, scattered among the cells filled with honey, or further down, we can see other cells, far fewer in number, which we can easily recognize by their colored, opaque contents. These are cells containing pollen (*p*, fig. 36). Generally, these cells are not capped.

26. Cells containing worker bee larvae. The worker brood

Now, let's examine the middle of the comb: there, we see certain cells that are also sealed off with a cap; but this cap is different from the ones we see on honey cells—it's a bit more dome-shaped (*s*, fig. 36).

---

* Bees probably do not add venom to their capped honey, as was believed before. Honey is preserved by its high sugar content, low moisture, low pH, as well as enzymes from bee stomachs that convert part of nectar into hydrogen peroxide. *Ed.*

If we open one of these cells, we find a young worker bee, still developing inside an extremely thin cocoon (*p*, fig. 37). At the bottom of the adjacent cells, which have yet to be sealed, we can see what look like tiny white worms (*y.l.*, *l*, fig. 36, and *l*, fig. 37). These are *larvae*, the first form a bee takes when an egg begins to develop. Finally, in still other cells, we can easily spot a small white egg affixed to the bottom (*e*, figs. 36 and 37).

Generally speaking, this entire section of the honeycomb used for laying eggs and raising young bees is called the *brood*, so called since

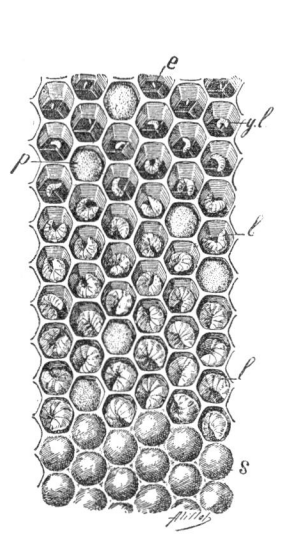

Figure 36. Piece of comb viewed from the front, showing cells containing pollen, *p*, and brood in various stages of development: *e*, egg; *y.l.* and *l*, larvae at various stages of development; *s*, sealed brood cells containing developing bees.

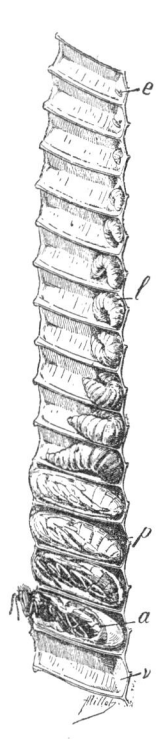

Figure 37. Lengthwise cross-section of comb showing bee development: *e*, egg; *l*, larvae at various stages of development; *p*, pupae in their cocoon; *a*, a bee tearing open its cocoon and emerging from the cell; *v*, empty cell after the bee has emerged.

CHAPTER 2. THE COLONY

the bees, much like mother hens, brood over these cells while the larvae are developing.

If we look at the brood area as a whole, we can clearly see the strict order according to which the mother bee lays her eggs in a honeycomb. Any empty cells in the middle of the comb are those from which fully developed bees have just emerged (*v*, fig. 37). Those that are covered by caps, domelike in shape to varying degrees, contain bees that have yet to reach their final stage; entomologists refer to bees in this state

Figure 38. Piece of comb showing both drone brood, *D*, and worker brood, *W* (photograph, 4/10 actual size).

of development as *pupae* (*p*, fig. 37). As a group, these cells are called the *sealed brood*; the surrounding cells contain larvae, and those furthest from the center contain freshly laid eggs. This tells us that the queen began laying her eggs from the center, moving outward as she went—and she'll only be able to resume laying eggs from the center when a sufficient number of empty cells are freed up there.

Figure 37 shows a lengthwise cross section of comb; the center of the brood is located at the bottom of the image, while the outer edge of the brood is nearer the top. Moving from the outer edge toward the center (from top to bottom in the image), we can trace the entire development of a worker bee, from egg to fully formed insect (fig. 37).

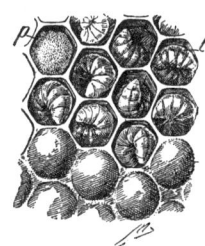

Figure 39. Piece of comb containing drone brood: *l*, drone larvae; sealed drone brood; *p*, pollen in a cell (actual size).

27. Drone cells. Drone brood

Now let's take a look at the drone cells (*D*, fig. 33), which are attached to the worker cells by a few irregularly shaped cells referred to as transition cells (*t*, fig. 33). In these cells, we again find everything we've just described. There may be drone cells containing honey or drone brood (*l*, fig. 39), and even, on occasion, pollen (*p*, fig. 39). Due to the drone's larger body, these drone cells are more pronouncedly dome-shaped and extend further outward than those of the worker brood (fig. 38).

28. Queen cells

During the swarming season (§ 39), we can find some cells in the comb that are quite different from those we've already seen, resembling whole peanuts, bulging outward and hanging down (*q*, fig. 33). They're covered with a network of half-constructed cells and end in a kind of small, bowl-shaped hollow. These special cells, which appear as if grafted onto the comb, are the *queen cells*.

By examining these cells at various stages of development (*c*, fig. 33), we can form some idea of how the bees construct them. In place of a worker cell and the cells surrounding it, the bees form a small cup that they then lengthen, enlarging it unevenly, so as to produce a hanging mass. An egg that looks much like that of a worker bee, but that will eventually become a queen, is stuck to the bottom of this cup. As the larva that emerges from this egg grows, the bees continue to lengthen the cell, which gradually takes on its final shape; eventually they seal it off.

29. How bees build comb

Using an observation hive, we watch the bees as they begin to construct their comb. As we've mentioned, the wax glands on the underside of their abdomen produce small flakes of wax; a worker building comb will detach these thin scales and hold them with her hind legs, then use her other legs to pass them up to the mandibles. There, they are kneaded into tiny balls that the bee applies, one by one, to various points where wax is needed to form the cells.

In this fashion, the workers begin to draw their comb from the top, roughing out the first cells, then building the comb out to the right and left and working their way down, such that the group of cells under construction takes on the shape of an oblong oval (fig. 40). At first, the cells of this unfinished comb are almost complete near the middle, but much shallower near the edges.

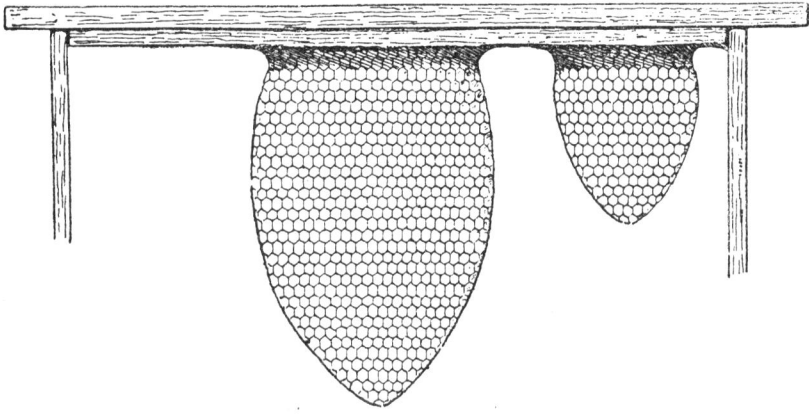

Figure 40. Initial comb construction.

30. New and old comb

Wax is white when first secreted by the wax glands, and so is the comb freshly built with that wax.

New comb is quite fragile; later, it takes on a yellow hue and grows harder. Moreover, the thin cocoons of the brood cells, collecting in layers one atop the other, thicken the walls of the cells and make them stronger. When comb is very old, it becomes brown or blackish, and at this stage it's quite hard and more resistant—and heavier.

31. Division of labor among bees

As we've seen, there are three kinds of bees in the hive:
1. An egg-laying queen (*Q*, fig. 43).
2. A large number of worker bees (*W*, fig. 41)—from 10,000 to 100,000 depending on hive strength.
3. A much smaller number of drones (*D*, fig. 42)—several thousand.

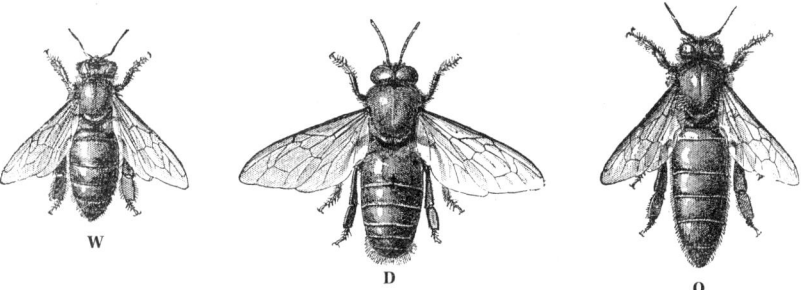

Figures 41 to 43. The three kinds of bees in a colony: *W*, worker; *D*, drone, or male bee; *Q*, queen (1/3 larger than actual size).

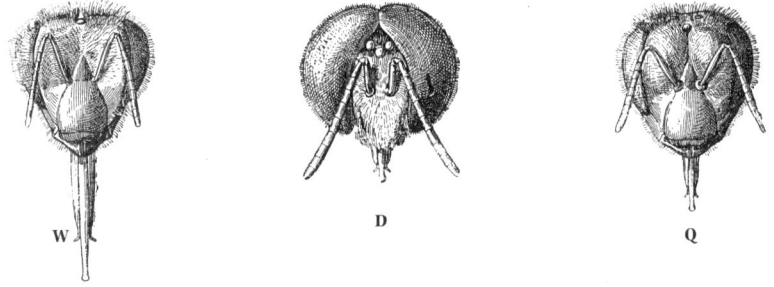

Figures 44 to 46. Heads of the three kinds of bees: *W*, worker; *D*, drone; *Q*, queen (5 times magnification).

We've already watched the worker bees as they go about their various tasks: guarding the entrance, ventilating the hive, foraging, storing up provisions, raising the young, or building with wax.

You might think that there are several kinds of workers—some dedicated to working with wax, some to harvesting, standing guard, providing ventilation, raising the young, etc. But this isn't the case: all of these tasks can be carried out by the same bee at various ages.

When a young bee first emerges from the cell after maturation, the workers clean it up, brush it and give it some honey to eat.

This young bee is still too weak to go out foraging. Instead, it stays busy with various jobs inside the hive: it uses honey and water to make a nutritious jelly that is fed to developing larvae, or it joins in on comb construction, if any is underway.

Later on, the worker bee begins to leave the hive. You may spot such young bees on their first time outside, as they learn to fly, and to recognize the objects surrounding the hive, as well as the hive itself.

We've mentioned that on a nice day you may quite often notice these recently-emerged bees moving away from and back toward the hive, flying in circles of varying diameters (§ 11).

When the bee has gotten used to going out, to recognize her hive and find her way around outdoors, she first spends most of her time gathering water, and later pollen and nectar.

When the bee has grown too old for the tiring work of a forager—as seen by her fringed and worn-out wings and her almost hairless body (fig. 47)—she can still be of use to the colony

Figure 47. Aged worker bees (1/2 larger than actual size).

for a certain length of time, remaining inside the hive and helping to maintain the proper temperature.

For that matter, any middle-aged bee can serve as a guard, a fanning bee, a cleaner, etc., depending on the colony's needs.

Everything we've just observed shows us how remarkably a bee colony applies the principle of division of labor—from the queen, dedicated solely to laying eggs, to the workers, who, depending on age and circumstances, play highly varied roles in the life of the hive.

The drones, on the other hand, will never be seen working at all, whether inside or outside the hive: their only role is to fertilize virgin queens from other hives. For the most part, drones are nothing but extra mouths to feed; later, we'll see how a beekeeper tries to limit their number.*

When the warm season is over, the bees themselves get rid of the drones, who are now of no use whatsoever to the hive: they are denied entry to the hive, chased away, or mercilessly killed.

32. The lifespan of a bee

The three kinds of bees we've seen have varying life expectancies.

As we've just mentioned, the drones tend to disappear at the end of summer; the queen won't resume laying eggs to produce more drones until the following spring.

The life expectancy of a worker bee varies considerably, depending on the season and the kind of work they do.

In order to determine a bee's lifespan, the queen bee of a given hive was replaced by another, of a different race; the colony was monitored to see how much time passed until not a single ordinary bee remained in the hive—that is, how long it took for all the ordinary workers to be replaced by worker bees of the different race. The results showed that during the peak harvest season, workers live no longer than six to ten weeks. In spring and fall, they may live longer; in winter, when their activity level is extremely low, they tend to live even longer—but never as long as six months.

During sudden downpours or strong winds that arise while the foragers are out among the flowers, many workers may die outdoors. In one case, a colony was found to have lost as many as 4,000 bees during a storm.

The queen, on the other hand, can live as long as four or five years. She will live longer in a smaller hive, where her egg-laying is more

---

* Drones are integral to the genetic success of the colony. It is now increasingly recognized that bees should be allowed to rear a sufficient number of drones. *Ed.*

limited, than in a larger one, where she has almost unlimited room to lay eggs.*

33. Egg-laying

As we will see, when a queen bee is replaced—as, for example, when a former queen has left the hive with a swarm to found a new colony (§ 39)—then the few queen cells in the hive will yield several young queens, of which only one will survive, the others having been killed by the workers or even by the new queen herself.

First, the virgin queen remains inside for five to seven days, without laying eggs or leaving the hive; it is usually on the sixth day that she exits the hive to be fertilized outdoors. Then, she returns. Generally, she begins laying eggs on the eleventh day after emerging from her cell.

We've already seen the order in which the queen lays eggs across a comb (§ 26), but she doesn't wait until one comb is full before moving on to another. After a partial egg-laying on one side of a comb (fig. 48), she will pass to the other side to lay a certain number of eggs, then to another comb, where she'll again lay eggs in a certain number of cells, and so on.

Figure 48. A queen laying eggs.

Let's look at a hive, during springtime, where all of the combs are parallel to each other and of the same size—in a moveable-frame hive, for example. We notice that one comb, comb A (fig. 49), located in the middle of the group formed by the bees—that is, in the middle of the hive's brood nest, holds the widest circle of brood on its

---

* In a queenless hive, one may sometimes still see eggs. These eggs are laid by workers who appear no different from the others, and who are referred to as *egg-laying workers*. The eggs laid by these workers can be found in drone cells or in worker cells, but in any case they are drone eggs, meaning that the colony is still doomed despite the efforts of the laying workers, since no more workers can be produced.

In laying-worker hives, queen cells are sometimes constructed, but any egg laid there by the workers never develops.

two sides; meanwhile, the two combs *B* and *C*, situated to the left and right of comb *A*, have smaller circles of brood, while the two combs *D* and *E*, to the left and right of the combs we've just mentioned—*B*, *A*, and *C*—have even smaller circles.

Whether, as in this example, the brood is distributed among just five combs, or whether it includes a larger number, its general distribution is roughly the same: oval in shape, and longer horizontally, perpendicular to the plane of the combs (figs. 49–51).

On the basis of this knowledge, it is easy to see how the queen goes about laying her eggs. Moving, as we said, from one comb to another, she lays her eggs about the circumference of all the circles of brood, so as to maintain uniform development across the multiple combs—such that the older brood, *o* (in 1, fig. 49), is found near the center and the newer brood, *y*, near the exterior. After a certain time, the brood in the center matures and the bees emerge, leaving behind empty cells, *v* (in 2, fig. 50). As the young bees gradually leave the

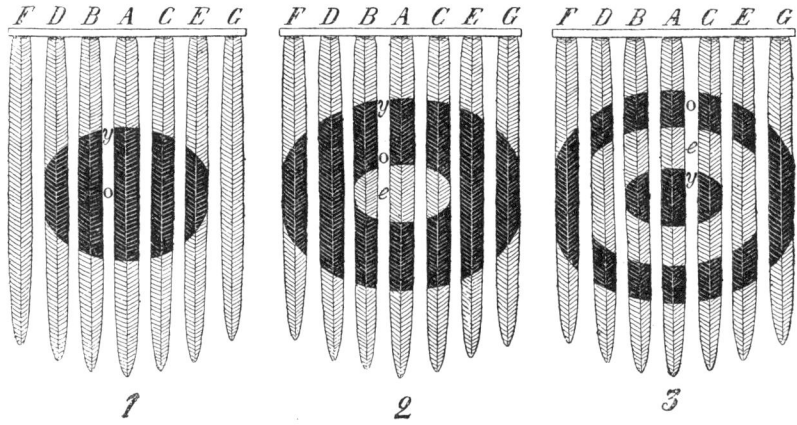

Figures 49 to 51. Theoretical diagrams showing egg-laying patterns and brood development. *F, D, B, A, C, E, G*, comb cross-sections. In *1*, the queen has begun by laying eggs in the center: *o*, older brood; *y*, young brood; the queen proceeds to lay more eggs outside of *y*. In *2*, the brood in the center has emerged, and the cells, *e*, are now empty; *o* and *y*, older and younger brood, respectively; the queen proceeds to lay more eggs outside of *y*. In *3*, the queen has begun once again to lay eggs in the center; *y*, young brood; *e*, cells that are empty after bees emerged; *o*, older brood; the queen continues laying eggs outside of *y*, in the cells marked *e*.

cells in which they were formed, many cells stand empty near the middle of each brood circle; the queen resumes laying her eggs in the vacated cells, starting with the middle of the largest circle, which is on the central comb, *A*. The queen continues to lay on the other combs in such a way as to ensure that the overall distribution of the new brood forms a smaller oval, *y*, that is contained within the first oval, *o* (in 3, fig. 51). As the fully developed bees depart, and the outer brood cells are left empty, the new brood, produced within the first one and constituting a smaller oval, expands gradually until it occupies the same volume as before; the queen again begins laying her eggs in the middle; and the cycle repeats itself.

We should add that the brood needs constantly refreshed air in order to develop, as it breathes intensely. For this reason, one always finds the brood on the combs nearest the hive entrance, which is the source of the ventilation.

Depending on the intensity of the queen's egg-laying, the oval-shaped brood mass will vary in size—and its volume generally serves as a means for judging the strength and well-being of the colony.

34. The number of eggs a queen can lay daily

The number of eggs a queen can lay in a single 24-hour period can vary greatly. This quantity (between 0 and 4,000, or even more) depends primarily on four different factors:
1. The season and the honey flow.
2. The queen's age.
3. The space available to the queen for laying eggs.
4. The number of worker bees in the hive.

Let's consider the various causes of differing rates of egg-laying, one by one.

1. *Egg-laying depends on the season and the honey flow, or the amount of nectar coming in.* Generally speaking, the more actively the bees forage, the greater the number of eggs the queen lays. So in winter, when the bees don't leave the hive, the queen stops laying eggs almost entirely, while at the peak of harvest season she lays the greatest number of eggs per day—as long as there is enough room to lay them.

2. *Egg-laying depends on the age of the queen.* Generally speaking, the queen lays the most eggs during the first two years of her life; queens who are four or five years of age are much less fertile.

Of course, it goes without saying that the number of eggs two given queens of the same age can lay often varies considerably. Some queens are highly fertile, while others aren't—and since the colony's prosperity and the amount of honey produced depend on the queen's fertility, this matter is of tremendous importance for apiculture.

At the same time, it must be said that in most cases a colony's queen is replaced naturally. Indeed, a queen will only initiate a swarm (§ 39) when queen cells are in place, and since it is the former queen who departs with the swarm, the hive will be left with a new queen.

If a colony does not produce swarms, or if we are speaking of a swarm that has just settled in a new hive, bees tend not to wait for the natural death of a queen before replacing her. When a queen is no longer sufficiently fertile, the workers take it upon themselves to replace her by making new queen cells. The old queen is eliminated and replaced by one of the new queens—a process referred to as *queen supersedure*.

3. *Egg-laying depends on the available space.* During the foraging season, and during the queen's most fertile age, the number of eggs she lays may be affected by a lack of available cells to lay them in. At the time of the main honey flow, if there is not enough room in cells near the top or sides of the hive to store the honey, the workers won't hesitate to store the nectar they've gathered in any available cells, even in those near the center of the combs, which in spring, as we have seen, are reserved for the brood.

As a result, the queen's egg-laying is unavoidably disrupted, taking place irregularly in whatever empty cells she is able to find here and there. If the hive is relatively small, egg-laying can grind to a halt because of the harvest—and this is the primary cause of swarming.

4. *Egg-laying depends on the number of workers in the hive.* Assuming that all of the factors mentioned above are ideal for egg-laying, it may still be limited by a fourth factor.

As one can imagine, a colony must be able to dedicate a sufficient number of workers to the job of tending the brood. If they are too few, the queen will be forced to limit her egg-laying, even though she may be perfectly fertile and have plenty of available room. If, on the other hand, there is a good supply of workers, the queen will intensify her egg-laying as much as her fertility will allow.

So we can say that if all the other conditions for egg-laying are favorable, it will be in direct proportion to the hive's strength.

35. A drone-laying queen

During the foraging season, in addition to worker eggs, we know that there are drone eggs, located in the largest cells. These eggs are also laid by the queen, and she can even lay them without being fertilized.* In certain cases, the queen will not leave the colony, despite being unfertilized; in this case, she will only lay male eggs. Such a queen is referred to as a *drone-laying queen*—and her hive is doomed.

These colonies will retain their males during the winter. Drone brood can be found scattered haphazardly in worker cells, but their caps will bulge outward more noticeably than usual (*D*, fig. 52). These cells will produce drones that are smaller in size, but perfectly normal otherwise. Figure 52 shows a section of comb where one can see both worker brood, marked *W*, and drone brood, marked *D*, in worker cells.

36. Development of a worker bee

As we have said, bees in the form of eggs or larvae, and bees that have yet to emerge from their cells, are referred to collectively as "brood."

Let's take a look at how the brood develops, from the moment the egg is laid until the fully-formed insect emerges from its cell.

Three days after an egg is laid (*e*, fig. 37), it has transformed into a kind of small legless worm. This is a young larva (*l*, fig. 37), which the workers now begin to feed by placing at the bottom of the cell a kind of jelly which they make in their stomach by mixing water, honey, and pollen.

---

* This is the phenomenon referred to as *parthenogenesis*.

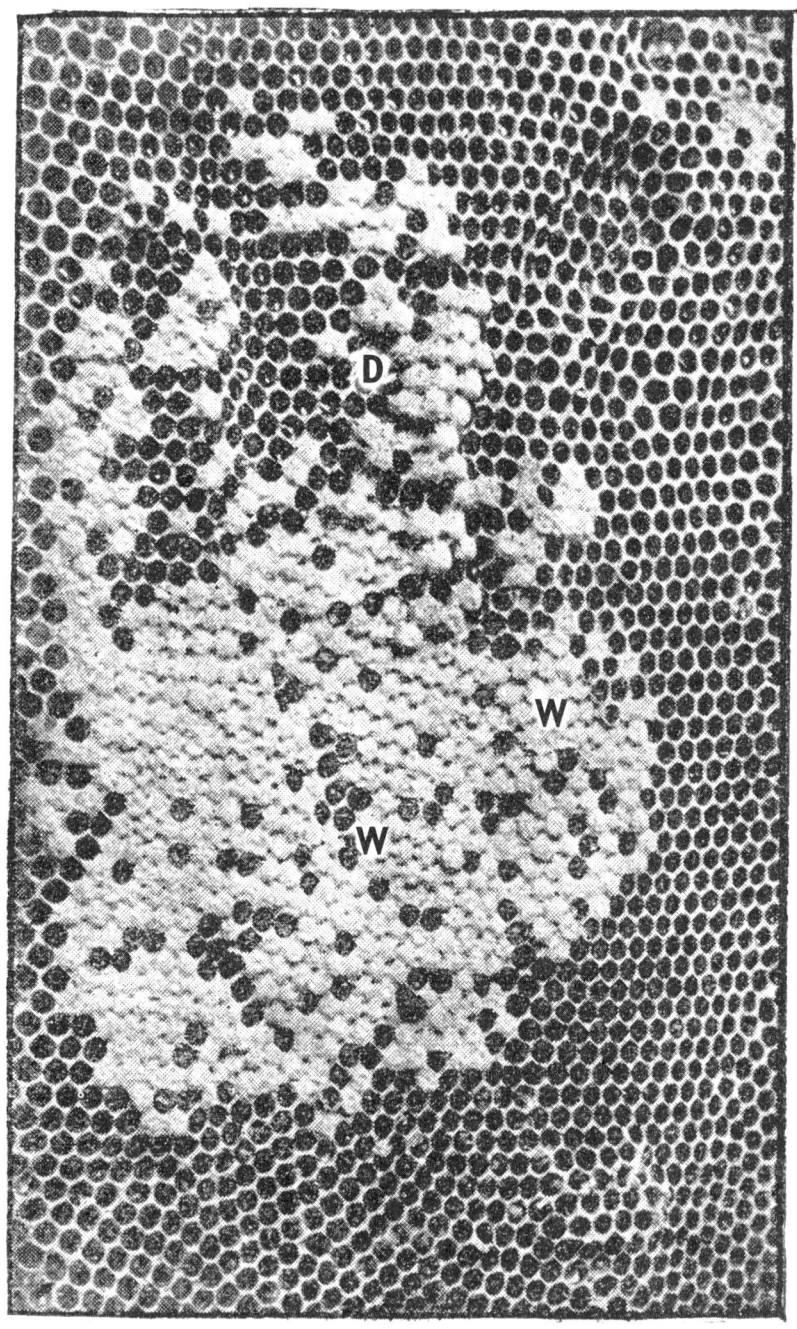

Figure 52. A piece of comb, showing worker brood, *W*, and drone brood, *D*, in worker cells (.45 actual size, photograph).

The larva—tiny at first (*y.l.*, fig. 36) and floating in the middle of this jelly—grows very rapidly, becoming longer, filling the length of the cell at the end of five days (eight days after the egg was laid) and storing up an abundant reserve of food in its body. It is at this moment that the bees seal the cell with a cap. At this point, they have nothing left to do but maintain, by their presence, the warmth necessary for the young bee to develop.

Next, the larva spins an extremely thin cocoon that surrounds it completely; it then molts and turns into what is called a *nymph*, or *pupa* (*p*, fig. 37). Next, from the eleventh day on, the pupa is gradually transformed into an adult, without any noticeable change in volume: the three parts of its body become more distinct, and at the same time one can see something resembling nipples taking shape under the head, which will later grow into legs. Finally, on the twenty-first day, the adult bee punctures the cap itself in order to emerge from the cell (*a*, fig. 37). The abandoned cell (like that marked *v*, fig. 37) must now be cleaned by the workers before the queen can lay another egg there.

37. Development of a queen bee

A queen develops in roughly the same manner, except that the bees provide the larva with a special kind of food whose consistency and taste differ from those of the food given to workers. In addition, a queen's maturation period is different.

The queen cell is capped after the same number of days have passed, but the transformation into an adult insect goes more quickly in this cell, lasting roughly seven or eight days, giving a total of fifteen to sixteen days from the moment the egg is laid until the queen's development is complete.

38. Development of a drone

Drones develop much as workers do, but take a bit longer. The cell isn't sealed until seven or eight days after the egg is laid, and the development period lasts a total of twenty-four days.

39. Swarming

Up to now, we've examined how a bee colony is organized, and we've seen how its life is tied entirely to that of the queen. If she dies and isn't replaced in time, or if she hasn't been fertilized and goes on to produce only drones, then her entire family is destined to die out.

So a bee colony can be thought of as one whole, a living being that can die much like a single organism would.

But just as a colony can die, a colony can be born: bee societies, each constituting a single complex organism, can multiply and reproduce, just as isolated organisms can.

The process by which colonies multiply is called *swarming*, and an emerging new colony is called a *swarm*.

Swarms are most common in early summer, when, as egg-laying and foraging simultaneously intensify, the population outgrows its hive.

When a hive is on the verge of casting a swarm, there is always a certain number of queen cells in progress. Five or six days before the most mature queen cells are expected to hatch, the existing queen leaves the hive, accompanied by a certain percentage of the hive's population. The colony that produced the swarm is now left with the remaining bees, and, five or six days later, will have a young virgin queen—and only one, since the others will have been killed by the queen herself or by the other bees.

As a result, a single colony has now become two:
1. The swarm, led by the *old queen*, which seeks to establish itself somewhere else.
2. The existing colony, now with a smaller population and a *new virgin queen*.

40. A swarm is cast

A few signs have been cited to presage the emergence of a swarm, such as the appearance of a large number of drones (which coincides with the production of queen cells), a population that beards outside the hive (§ 12), and even the coming-and-going of numerous worker bees, from the hive's interior onto the landing board or vice versa.

But none of these signs tells us anything for certain, especially since a swarm's emergence depends on the weather and the outside temperature.

Swarms are rarely seen when the temperature is below 68°F (20°C), or when flowers are producing little nectar. Generally, swarms emerge around midday, between 10 a.m. and 3 p.m.

The swarming season varies based on the climate and the local honey plants. In our temperate regions, swarming most often occurs in May and June. In the Mediterranean region, it occurs in April and May. In the mountains, it's even later, in June and July. Finally, in regions where buckwheat and heather grow, swarming may occur as late as August.

Figure 53. A swarm hanging from a tree branch.

When a swarm is cast, an enormous mass of bees pours out of the hive, circling it or flying in all directions while rising in the air. But after a very short time, as if answering a rallying cry, they reunite at a single spot—on a tree branch, from which they hang suspended one from another in a compact mass (fig. 53), or in a bush, beneath a beam, or even on the side of a wall. Sometimes they gather in a hollow tree trunk, a chimney, or any other cavity they come across. In the latter instance, one will have noticed worker bees which, before the swarm emerges, have traveled here and there, scouting the environs for a favorable location where the new colony can install itself.

Let's consider the most frequent case—when the bees reassemble beneath a tree branch. Once there, in this interim state, the swarm awaits the moment when it can hope to find some more permanent shelter and begin building there. Quite often, the swarm will only remain attached to the branch until the following day, when it departs for some more distant location, until it has chosen somewhere to settle down once and for all. It may happen that, for lack of any appropriate spot, the swarm continues moving further and further, losing a certain number of bees by the day, growing smaller and smaller until it finally dissipates.

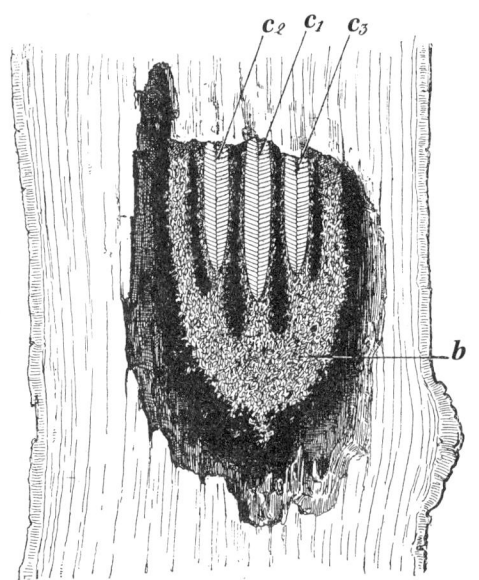

Figure 54. A swarm beginning to build comb in a hollow tree trunk: c_1, c_2, c_3, first combs constructed; b, clustered bees, hanging from one another (a lengthwise cross-section of the tree and the swarm).

When the swarm has found a suitable shelter, it immediately sets about building comb (c_1, c_2, c_3, fig. 54). In this respect, we should note that the workers who form the swarm have gorged on honey, and the majority of them are secreting wax in abundance. The new colony is now established, and becomes a wild hive.

A recently departed swarm may also return to the hive, either because the weather has taken a sudden turn for the worse, or because it has lost the queen.

41. Primary swarms and afterswarms. The queen's song

If the population left behind in the hive following the swarm's departure is still relatively large with respect to the capacity of the hive, then yet another swarm may emerge, called a *secondary swarm* or *afterswarm*. As we've said, the first young queen doesn't emerge from a queen cell until five or six days after the primary swarm. If an afterswarm is in the works, then the remaining queens, still in their cells, are not killed, and for a period of one to three days the young queen sings a peculiar song, which sounds something like *tih, tih, tih*, and is easy to hear in the evening. The queens who are still in their cells respond to this song with one of their own: *kwa, kwa, kwa*.

These special songs—extremely easy to recognize—warn the beekeeper that an afterswarm is about to emerge from his hive.

So if the weather is favorable, the afterswarm emerges approximately eight days after the primary swarm. Once the afterswarm has departed, the remaining bees, which have left the other queens fully developed in their cells, allow one to exit and kill the others.

On occasion, however, it may happen that the other queens are held prisoner for a while longer; in this case, the second virgin queen sings *tih, tih, tih*, like the first virgin queen—which indicates that a *tertiary swarm* may emerge a few days after the secondary swarm.

Still, it must be said that the emergence of swarms depends on the weather and the outside temperature. The queens' songs indicating a coming afterswarm are never a sure indicator that these swarms will actually emerge. If the weather suddenly deteriorates, then the swarm will not emerge, and the young, imprisoned queens will be killed.

Note: We can tell that a queen is about to hatch by the fact that the cap covering its cell has begun to be gnawed at about the end, which opens when the queen emerges. When the bees kill a queen in her cell, the queen cell is opened from the side.

Summary

The colony
During the warm season, a bee colony includes: 1) a queen; 2) a great number of workers; 3) a much smaller number of drones.

1. The *queen* lays all of the colony's eggs, and lives for several years.

2. The *workers* perform all of the work in the hive; during the honey flow, they live between six weeks and two months.

Inside the hive, their jobs include: building comb with the wax secreted in scales from their wax glands; stocking the comb with honey and pollen; raising the brood—that is, the young bees, from egg to adult insect; building, when necessary, queen cells and raising replacement queens; ventilating the hive during the harvest to eliminate the excess water in the new honey; and finally, cleaning the hive, guarding the entrance, etc.

3. The *drones* don't work; their only role is to fertilize the virgin queens from other hives. In autumn, they are either driven out of the hive or killed by the worker bees.

Colony development. Swarming
In winter, there are usually no drones in bee colonies; the queen has stopped laying eggs completely, or almost completely; and the workers, gathered around her, never leave the hive. Without working, the colony feeds on the honey stockpiled in the comb during the preceding foraging season.

During the first nice days of spring, the queen bee resumes laying her eggs; the workers venture outdoors to look for nectar and pollen on the first flowers; they also gather water and raise more and more brood, growing in the hive. Bit by bit, the number of bees rises; the brood nest is expanded considerably, and, when the strong honey flow begins, all of the hive's comb may become filled with honey and brood. During this period, we can also see some drone brood appear, and the bees often build cells for replacement queens.

If the hive has now become too small for the colony, a portion of the bees will leave the hive with the existing queen; this is a *primary swarm*, which will establish a new colony.

A hive that has produced a swarm has a new virgin queen, who may leave the hive herself eight days or so later, with an *afterswarm*; there may even be additional swarms. In any case, the original hive is left with a new queen.

In autumn, the workers evict or kill the drones, and the brood gradually diminishes.

When winter begins, the bees gather around the new queen, who now lays much fewer eggs or none at all; they spend the cold season within reach of their honey stockpiles—and the cycle repeats itself when spring arrives.

As for the swarms that have left the hive, they spend the winter just like the original hive, having built new combs and filled them with honey and brood.

Chapter 3
The Hive

42. Fixed-comb hives

Figure 55. An apiary with skeps.

Wild honey bees usually establish their colonies in old, hollow tree trunks. So the first humans to keep bees probably had the idea to install a swarm in a natural nest.

The first artificial hive was probably a tree trunk that had been hollowed out, sawed off on the top and bottom, and covered with a board or flat stone (fig. 56). Such hives are still to be found in certain regions (see fig. 132).

Wherever the cork oak grows, keepers prefer to use its thick, waterproof, and easily detachable bark to create the outer casing of a bee home. The result is already more elaborate than a simple tree trunk.

In other regions, especially in mountainous ones, hives that are taller than they are wide are built by simply assembling four boards and nailing a cover on top (fig. 57). This is still a primitive hive, but with a more regular form.

In many other areas, keepers have housed their bees in a rounded basket, often pointed on top, sometimes low and squat in shape, and made either with straw or with wicker interwoven in a regular pattern. Each hive is covered with a *hackle* (hood) made of straw to protect against rain and temperature changes. This is the most common form of hive, called a *skep*.

All of these hives, from the simple and primitive tree trunk to the most meticulously built wicker hive, are referred to as *fixed-comb hives*.*

Figure 56. A log hive.   Figure 57. A hive made of planks.

In order to provide additional support for combs, wooden sticks are typically placed across the hive interior.

To help protect the bees against predators, hives are generally set atop a support made of wooden boards, called a stand, elevated off the ground in one way or another (figs. 55, 57 and 58).

Since bees store their honey reserves to the top and sides of their brood cluster, we can easily grasp the primary advantages and dis-

---

* Despite thousands of years of successful use, fixed-comb hives are now outlawed in some jurisdictions. *Ed.*

advantages of the various kinds of fixed-comb hives we've just mentioned. Hives made of tree trunks, cork or boards have a top that is removable, like a lid, making it possible to harvest the honey located in the upper portion of the hive without excessively disturbing the bees. Bell-shaped skeps made of straw (fig. 59) or wicker (fig. 58) are shaped so as to allow the cluster of bees to consume their honey reserves gradually during the winter. At the end of autumn, the cluster

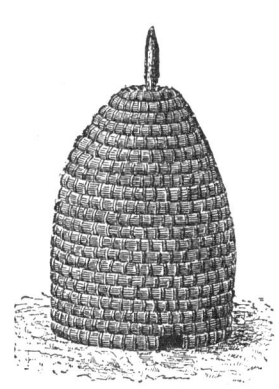

Figure 58. A skep made of wicker, with a straw hackle.

Figure 59. A skep made of woven straw without its straw hackle.

is situated near the bottom of the reserves, and then begins moving upward and consuming the honey bit by bit. Rising through the hive as the cold season passes, the cluster is always near the honey it needs, thanks to the shape of the hive. But, aside from this advantage, these hives make harvesting honey inconvenient. If you want to avoid killing the bees, and lack knowledge of appropriate beekeeping methods (driving bees into a different hive prior to harvest, changing hives places, artificial swarms, etc.), you'll have no choice but harvest honey from the active hive by turning it upside down and cutting out the comb.*

Fixed-comb hives made of woven straw or wicker—the skeps—are covered with a straw hackle (fig. 58) that protects them against rain and cold.

43. Suffocating bees

Since cutting out honeycomb is an operation requiring a certain amount of experience with bees, many beekeepers find it much simpler to burn a sulfur wick beneath the colony, thus killing all the bees in order to sell the contents of their hive. This is called *suffocation*. This deplorable but all-too-common practice isn't even justified by any real advantage, since by killing the bees the beekeeper is losing a significant share of his capital.

44. Cap hives

Keepers have sought to combine the advantages of the various kinds of fixed-comb hives and avoid suffocating the bees by building hives consisting of two parts, stacked atop one another, and made of boards or straw cords. These are called *cap hives*.

The cap is like a second, smaller hive set on top of the first one; if there is only one cap, its volume should be adjusted, depending on the region, so as to contain only the surplus portion of the honey reserves. In this case, the ease of harvesting the honey filling this cap, without disturbing the bees, is clear. Generally, this hive system is superior to the two described above, and making it is hardly more complicated.

Here, for example, is a description of a good cap hive, made of wood, that is suitable for a majority of cases (fig. 60); it can also be made of straw cords (fig. 61).

The lower portion, or the *hive body* (*B*, fig. 60) should have a volume of between 2,500 and 3,000 cubic inches (10–13 gal, 40–50 L), in the form of a wooden case, made of planks 1 1/4" (3 cm) thick.

To allow the bees to build their combs parallel to each other, the ceiling of the hive body is made with slats called top bars (*t*, fig. 60).

---

* Chapter 15, from § 195 on, contains more information on caring for fixed-comb hives.

Based on what we've learned by examining how bees construct their comb, these top bars should be made and arranged as follows: each slat is 1 1/8" (28 mm) wide and 3/8" (1 cm) thick, and the centers of two adjacent top bars are spaced at 1 1/2" (38 mm) apart, leaving 3/8" (1 cm) between them. To force the bees to build following the direction of the top bars, it is best to use hide glue to attach strips of old comb from a dead hive to the bottom of the top bars (see for example the upper portion of fig. 70). The cover has a large opening, above which the cap can be placed. If a cap is not being used, the opening can be covered with a board.

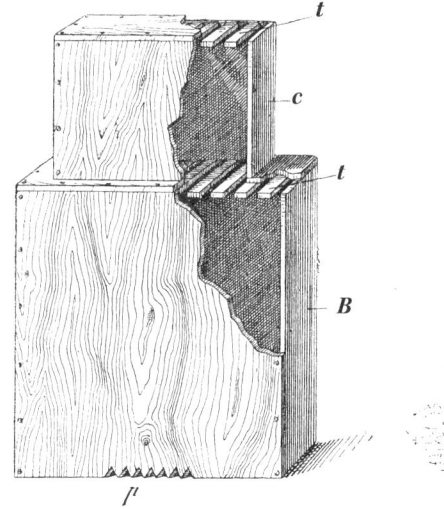

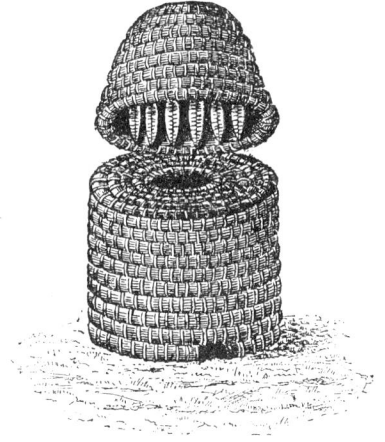

Figure 60. A cap hive made of wood: B, the hive body; c, the cap; t, top bars to guide construction of the comb.

Figure 61. A cap hive made of woven straw.

During peak honey flow, this board can be removed and replaced with the cap.

The *cap* (c, fig. 60) is like a second, smaller hive, and should have a capacity of 900–1,500 cubic inches (4–6.5 gal, 15–25 L). It's a good idea to glue some comb beneath the top bars in the cap to lure the bees up into it. When the cap is full of honey, it is removed and replaced by another if the flow is still ongoing. At season's end, it is removed for good and replaced by a board—leaving only the hive body throughout the winter season.

There are many other models besides the one just described—for example, those whose hive bodies have a simple hole on top (fig. 61), through which the bees climb up into the cap once the hole is unplugged.

However, it has been observed that the bees are less willing to climb into the cap through this hole-shaped opening than through the gaps between the slats described above.*

45. Stacked hives

In the hive body of a cap hive, the comb is rarely renewed, and, after a certain time, the aging comb can hinder the bees' development. Already long ago, in order to encourage comb renewal, keepers conceived the following more complicated hive system.

The hive is divided into several segments of equal size, stacked atop one another; they can be made of wood or of straw cord. Each of these segments is called a *super*.**

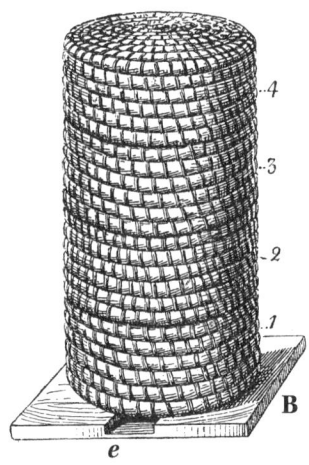

Figure 62. A stacked hive made of woven straw: *B*, bottom board; *e*, entrance; 1–4, stacked supers.

Figure 62 shows, for example, how a straw hive consisting of four supers is constructed. As one might guess, when the upper super (which serves as the cap) is harvested, one adds another underneath where the bees can build new comb.

We cannot really recommend using these old hives—their inconveniences outweigh their advantages.

Splitting up the hive into multiple supers aims to make certain tasks easier for the beekeeper, who can expand or contract the volume

---

* See §§ 207 and 208 concerning how to manage cap hives.

** One shouldn't confuse this system, where all the supers are identical, with fixed-comb hives to which a small super may be added on top, or with vertical frame hives (§ 171), which may also be referred to as hives with supers.

of the hive, or combine two relatively weak hives into a single stronger one. But using these hives is complicated. There are still other disadvantages: when a super is added to the bottom of the hive during the harvest, the bees often build a large number of drone cells there—and, as we have seen, a beekeeper should do everything possible to prevent the excessive construction of drone cells, which would lead to too many extra mouths to feed.*

Moreover, stacked hives are bad for winter, because their horizontal divisions hamper the movement of the bee cluster that is unable to winter in a single super. The horizontal gaps also hinder regular egg-laying by the queen.

As a matter of fact, based on our study of a bee colony, it's easy to see that anything that breaks up the unity of the brood nest must be harmful.

For all of these reasons, the old-time stacked hives are being used less and less these days. Although they seem to be an improvement, they actually produce poorer results than the cap hive or even the ordinary skep.

46. Movable-frame hives

No matter which hive one chooses from those considered above, its management will present quite a few difficulties for anyone wanting to maximize the honey harvest. With fixed-comb hives, a simple beekeeping technique will not be very productive, while any more productive technique will prove complicated.

We've seen how—using a cap hive, for example—one can encourage the bees to build their combs in regularly arranged rows using guide bars "primed" with old comb. In time, beekeepers wondered whether it would be possible to leave these bars unfastened, and prevent the bees from attaching their comb to the hive walls by using two other vertical bars joined to each horizontal one, such that each comb, built according to plan within the empty frame formed by the three bars, could be removed from the hive—that is, a *movable comb*.

---

* As noted in an earlier comment, we now know that a healthy colony needs an adequate number of drones. *Ed.*

A hive of this kind, called a *movable-frame hive*, was quite easy to make: it's really nothing more than a box of the cap hive described previously in which each guide top bar, spaced at the distances indicated above, is replaced by a frame, filled out with a final horizontal bar at the bottom (figs. 63, 64 and 65).

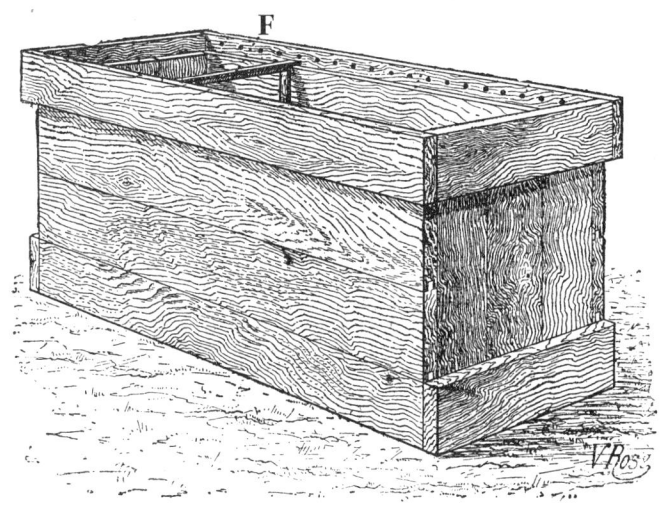

Figure 63. Exterior of a moveable-frame hive containing a frame, F.

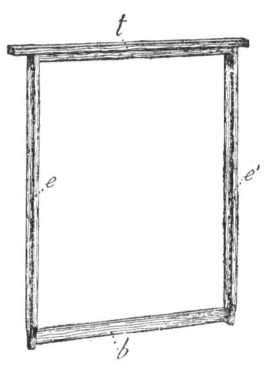

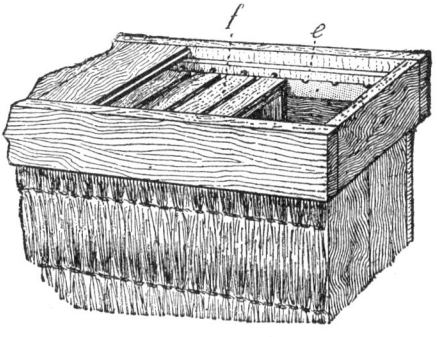

Figure 64. A frame from a moveable-frame hive: *t*, top bar; *e*, *e'*, end bars; *b*, bottom bar.

Figure 65. Fragment of another moveable-frame hive, viewed from above: *f*, frames in place, alongside one another; *e*, a section where frames have not yet been placed.

In short, a *frame hive* is simply a wooden case filled with wooden frames, arranged parallel to each other. This case can be topped with a lid of some kind (fig. 66). Such frame hives are called *horizontal hives* (see § 98).

47. The advantages of frame hives

Moveable-frame hives can be inspected very easily, since the beekeeper can remove each frame individually to examine it as he wishes (fig. 67) then put it back into place. But the primary advantage of these hives is that they make it easier to harvest part of the honey. Using a very simple tool called an *extractor* (fig. 68), the keeper can remove the honey from the cells while leaving the waxen combs intact, which are given back to the bees, who eventually refill them with honey.

To harvest honey in this way without breaking the comb, the beekeeper first removes a frame that is full of honey, and, using a special knife, trims off all of the caps of the cells on both sides of the comb.

When, for example, four frames have been prepared in this fashion, one can use centrifugal force to drive out the liquid honey without destroying the wax. For this purpose, the four frames, placed behind a wire mesh to help protect the comb, are spun rapidly around a vertical axis. This system spins inside a container that collects the honey, which is then drawn out through a tap at the base.*

The cost of an extractor may seem high if you only have a small number of hives, but it must be said that the price is really not so high, especially since several beekeepers can share a single extractor, much as a threshing machine is shared among several farmers.

The frame hive has other important advantages. To feed a colony whose stock is running low (§ 87), or to provide honey reserves to a hive insufficiently prepared for winter (§ 127), one simply has to add one or several frames containing honey taken from stronger hives. Moreover, since frame hives are easy to check on, a keeper can keep better track of how each colony is faring.

---

* Later in the book (§ 129), you'll find a more complete description of these procedures, and, in § 225, a description of various kinds of extractors.

Figure 66. A moveable-frame hive on its stand.

Figure 67. A beekeeper inspecting a moveable-frame hive.

Finally, if the bees are enjoying a strong honey flow, they will have a large number of combs, already built and waiting to be filled with honey—increasing the honey yield.

Although this hive system may at first glance seem more complicated than those described previously, it actually rewards the beekeeper with a greater yield than a fixed-comb hive, for the same amount of work. In fact, modern beekeeping methods became possible thanks to the movable-frame hive.

48. Wax foundation and its advantages

Here are the most important points for using movable frames successfully:

1. In order to easily remove any given frame from a movable-frame hive, the bees must build their comb inside the frames in a regular fashion.
2. As we've just seen, it's very useful to have a sufficient number of comb-filled frames ready to give to the bees, to be filled with honey during honey flow.
3. Finally, as we've said repeatedly, the beekeeper must avoid excessive construction of drone cells.*

Figure 68. Harvesting honey using an extractor.

When using movable-frame hives, you can attain, quickly and simultaneously, these three advantages, especially if one equips the frames ahead of time with wax *foundation sheets*. Each sheet, which

---

* Allowing the bees to raise a large number of drones will indeed decrease the honey crop. However, drones are essential for colony health, and many natural beekeepers let the bees determine how much drone brood they need. *Ed.*

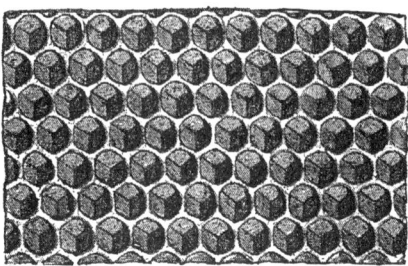

Figure 69. A sheet of wax foundation.

Figure 70. A frame primed at the top with bits of comb.

should be made of real beeswax, is stamped on both sides with a precise pattern of the bases of worker cells (fig. 69).

1. Since sheets of foundation are positioned perfectly straight inside the frames, when the bees draw out their cells the end result will be comb built regularly inside the frame.
2. Bees are able to rapidly transform foundation into finished comb that is ready for the harvest.
3. Since the cell pattern stamped onto the foundation consists completely of worker cells, the bees will naturally tend to draw out the cells according to these dimensions, rarely building many drone cells.

So the advantages of using foundation frames in beekeeping are obvious.

The expense of purchasing wax foundation is quickly compensated by the increased honey crop, especially since each frame, once purchased and built out with comb, can be used for a long time.*

Of course, anyone who would rather not incur this expense can, if necessary, do without foundation by:
1. Priming the top of each frame with small strips of old comb (fig. 70).
2. Waiting a bit longer for the frames to be built out by the bees.

---

* See also §§ 99 and 119 regarding wax foundation.

3. By cutting out, as needed, the portions of comb where the bees have built drone cells and replacing them with pieces of worker-cell comb taken from other frames.

Summary

Traditional fixed-comb hives
Wild honey bees typically live in the hollows of trees. The simplest kind of hive, called *fixed-comb hives*, include hollow logs covered with a board, wooden cases, and straw or wicker baskets (skeps).

Cap hives
A more advanced hive called a *cap hive* consists of two parts; the upper part, called a cap, can be removed, allowing surplus honey to be removed without disturbing the colony.

The bottom body of such hives is often outfitted with parallel wooden top bars, primed with strips of old comb to encourage the bees to build their comb in a more regular fashion.

Frame hives
If each of these top bars is replaced with a wooden frame inside which the bees build their comb, the frames can be taken out and put back at will. The result, called a *frame hive*, is a wooden case, inside which these movable frames, filled with comb, are arranged in parallel.

With this final kind of hive, honey can be harvested without harming the comb by using a special tool called an *extractor*, leaving undamaged comb ready for the bees to refill with honey. Another advantage of this hive is that frames can be taken from prospering colonies and given to weaker ones.

In sum, for the same amount of work on the beekeeper's part, frame hives yield considerably more honey than fixed-comb hives.

PART TWO
A BEEKEEPER'S APPRENTICESHIP

Chapter 4
A Region's Honey-Producing Potential

Figure 71. An apiary with moveable-frame hives.

49. Assessing local nectar plant resources

Generally, those wishing to take up beekeeping aren't free to choose which region they'll set up their hives in, since their other occupations tie them to a particular location.

However, any would-be beekeeper should study the honey potential of the given region before establishing an apiary there.

If, as is most often the case, the location of the colonies is dictated by the general layout of the property, one should examine the naturally occurring vegetation or crops that grow nearby, while also taking into account the climate and terrain.

Since bees will rarely stray more than a mile or two (2–3 km) from their hive, let's examine the resources they'll have at their disposal within this radius.

Here are a few essential practical tips concerning this all-important topic.

WILD NECTAR PLANTS

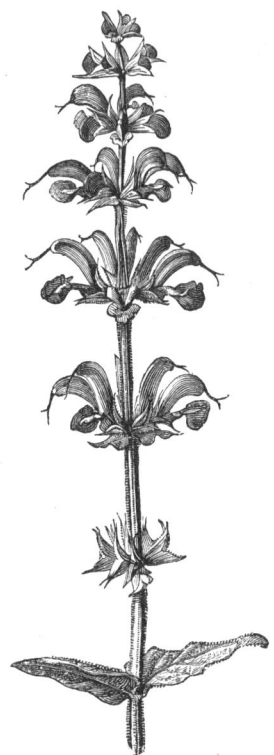

Figure 73. Knapweed (pink flowers, actual size).

Figure 72. Meadow sage (blue flowers, 1/2 actual size).

Figure 74. Sweet-clover (yellow or white flowers, actual size).

Figure 75. Wild thyme (pink flowers, actual size).

50. Wild nectar plants

If the only plants surrounding the potential hive location are naturally occurring vegetation, they will in most cases be plants of the meadow, forest or heath. If the hives are placed in an area dominated by natural meadows that are cut for hay, and not used for pasture, then the site is likely quite favorable for beekeeping, especially if the meadows feature such plants as white clover, meadow sage (*Salvia pratensis*, fig. 72), brown knapweed (*Centaurea jacea*, fig. 73), sweetclover (*Melilotus* spp., fig. 74), wild thyme (*Thymus serpyllum*, fig. 75), or the majority of meadow plants if one is in a mountainous region.

Fields used for pasture can also serve as an important resource, if such plants are highly abundant.

Figure 76. Bell heather, *Erica cinerea*, (pink flowers, actual size).

Figure 77. Common heather (pink flowers, actual size).

Figure 78. Goldenrod (yellow flowers, 1/2 actual size).

Figure 79. Blackberry (white flowers, 1/2 actual size).

Figure 80. Sycamore maple (greenish-yellow flowers, 1/2 actual size).

CHAPTER 4. NECTAR RESOURCES

Forests have the advantage of providing a flow throughout the season that almost always allows the bees to stock up their winter reserves, but this harvest often isn't particularly plentiful, and the honey is of mediocre quality. However, if there is a lot of heather in the woods (figs. 76 and 77), it can be a valuable resource in autumn. Still, it must be said that honey produced from heather is doubly inconvenient: aside from its inferior quality, it is too thick to be removed using an extractor.

In these regions, it's best to harvest honey before the heather begins to bloom, and to leave the heather for the bees as they stock up their winter reserves.

The primary resources in forests are plants found near the forest edge and in clearings (blackberry (*Rubus*, fig. 79), bugleweed (*Ajuga*), self-heal (*Prunella*), knapweeds (*Centaurea*), lungwort (*Pulmonaria*), goldenrod (*Solidago*, fig. 78), germanders (*Teucrium*), etc.) and certain trees or spring-blooming bushes like willow trees, wild cherry, wild plums, maple trees (fig. 80), etc. Another source of sugar-rich material for the bees to harvest in summer is the honeydew (§ 310) that falls in the form of a delicate rain from the leaves of many tree species.

Uncultivated areas, fallow fields, and railroad embankments are often covered by vegetation that includes such honey plants as blueweed (*Echium*, see fig. 1—the largest plant), thistles, toadflaxes, glastum (*Isatis tinctoria*), etc.*

In the Landes region of southwest France, various types of heather are the bees' greatest resource.

Prairies dominated by grasses, salt meadows, and marshes are among the least favorable areas with natural vegetation.

51. Nectar plants of cultivated fields and meadows

If the hives are located amidst fields of crops, it is very important to consider the nature of the plants being cultivated.

---

* For a more detailed understanding of the nectar plants in your region, you can refer to field guides to identify plants. In *Nouvelle Flore*, by G. Bonnier and de Layens (Paris, ed. Paul Dupont), the plants sought out by bees are specially marked.

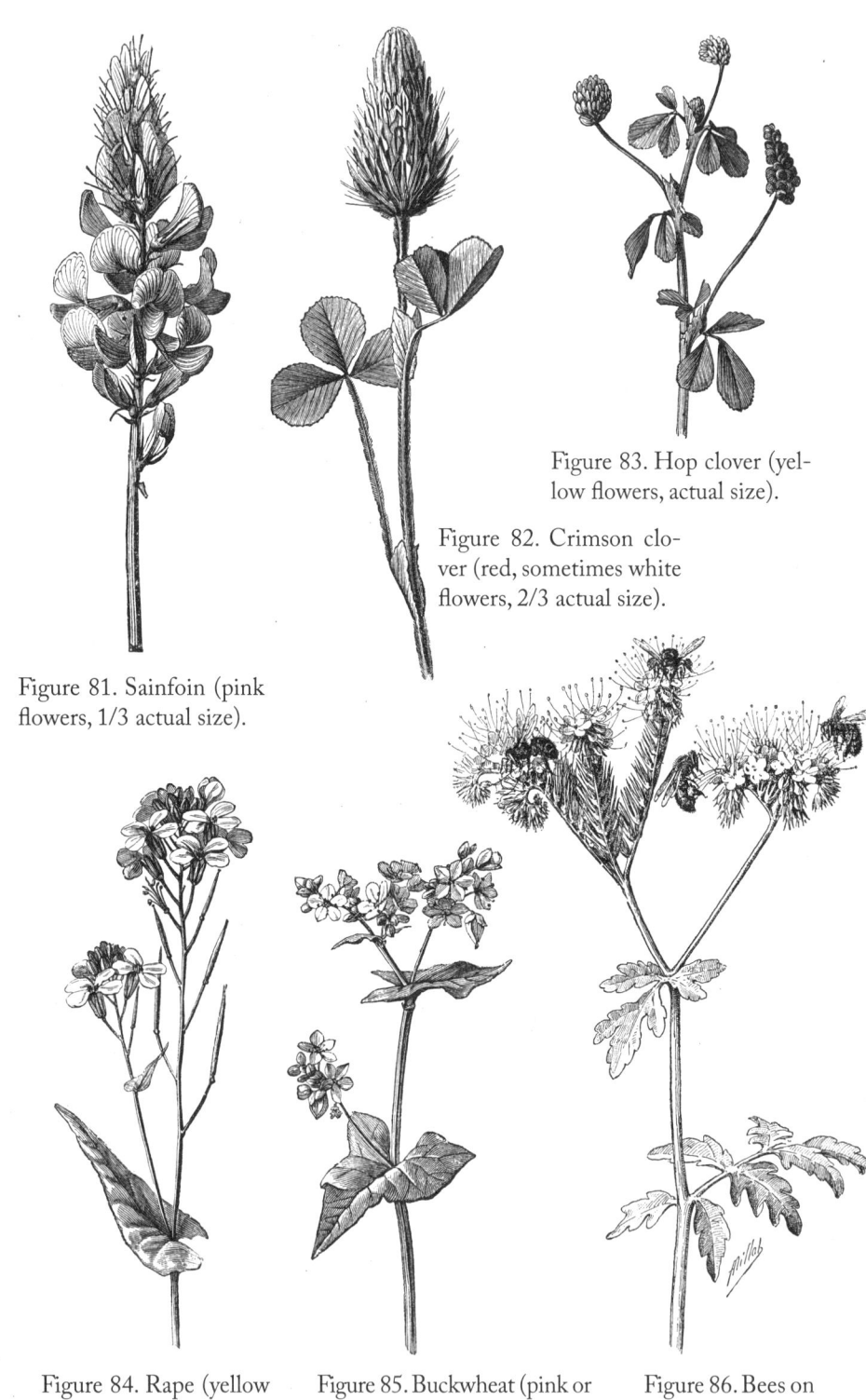

Figure 83. Hop clover (yellow flowers, actual size).

Figure 82. Crimson clover (red, sometimes white flowers, 2/3 actual size).

Figure 81. Sainfoin (pink flowers, 1/3 actual size).

Figure 84. Rape (yellow flowers, 2/3 actual size).

Figure 85. Buckwheat (pink or white flowers, 2/3 actual size).

Figure 86. Bees on *Phacelia* flowers.

CHAPTER 4. NECTAR RESOURCES

Without a doubt, the most favorable plant for bees is sainfoin (*Onobrychis*, fig. 81). One might go so far as to say that if an apiary is surrounded by sainfoin fields, a honey harvest is all but guaranteed, even in mediocre years. Sainfoin honey, known commercially as *Gâtinais honey*, is one of the most highly valued varieties.

Other notable forage plants include, most importantly: white clover (*Trifolium repens*, see fig. 21), crimson clover (*Trifolium incarnatum*, fig. 82), alsike clover (*Trifolium hybridum*), hop clover (*Trifolium campestre*, fig. 83), as well as vetches (*Vicia*) and sweet peas (*Lathyrus*). Alfalfa (*Medicago sativa*) only yields honey in the second cutting, and often has little nectar.

Next, we should mention the fields of rape (fig. 84), or even of cabbage, which can be an important resource in spring. In autumn, buckwheat (fig. 85) can produce plentiful nectar from which the bees produce a honey that, although mediocre in quality, is prized in the production of gingerbread.

Fields of wheat, rye, barley, oats, corn, beetroot, flax, and red clover* are essentially of no use in apiculture, unless poorly cultivated fields happen to include certain honey-producing weeds typically found on roadsides (cornflower, dandelion, wild vetch, delphinium, etc.).

Speaking of cultivated plants capable of producing honey, we shouldn't forget to mention beans and broad beans, peas, and flowering onions.

Once harvested, cultivated fields are sometimes invaded by other weeds that can also prove useful to bees: woundwort (*Stachys*), hemp-nettle (*Galeopsis*), knotweed (*Polygonum*, *Persicaria*), etc.

We won't discuss garden plants, since, despite their brilliant colors, they are in general rarely visited by bees. Nevertheless, we can note several garden plants that are honey plants: forget-me-nots, Alpine rock-cress (*Arabis alpina*), asters, *Phacelia* (fig. 86), etc.

52. Melliferous trees

Trees on roadsides, in forests, or along hedgerows, and even trees cultivated in fields or parks, can sometimes be an important source of

---

* Red clover flowers are too long for the bee's proboscis to reach the nectar, so bees only manage to harvest anything from them when the nectar is especially abundant.

honey. In the spring, these include, above all, willows and maples, as well as apricot, peach, cherry, and plum trees. Apple and pear trees, however, are rarely melliferous.

Later in the year, horse chestnuts, black locusts (fig. 87) and linden trees (fig. 88) become important, often yielding abundant amounts of honey.*

Figure 87. Black locust (white flowers, 1/2 actual size).

53. How favorable is a given region for beekeeping?

Speaking very generally, the aforementioned points lead to the following conclusions:
1. A region dominated by sainfoin and melliferous meadow plants (white clover), rape, and hop clover, and with linden and black

---

* The sweet liquid collected by bees from flowers was referred to as "honey" for over two millennia. This usage is preserved today in terms such as "honey flow" and even in the bee's scientific name, *Apis mellifera* (the honey-bearing bee). We chose to retain this traditional usage in many parts of this book, although *nectar* would be the more modern term to use. *Ed.*

locust trees present as well, will be an excellent region for apiculture, and the honey produced there will be of high quality.
2. A region with abundant buckwheat in the fields and with heather in the forests or moors can be quite good for beekeeping, but the honey will be of lower quality.
3. If the region is dominated by forests, bees will usually find resources sufficient for a modest harvest, but the region will rarely prove favorable for beekeeping.
4. A region with large beetroot, cereal grains, flax, or hemp crops or vineyards, but without melliferous trees or meadows, is poorly suited for apiculture, and will never yield a plentiful harvest.

Figure 88. Linden (yellowish flowers) (1/2 actual size).

54. The climate's influence on honey production

1. *Mountain climate.* The climate of France's mountainous regions is known to boost nectar production in plants, as well as the quality of the honey harvested.*

As long as the hives are not at too high an altitude, where the season is too short, it is safe to say that bees will always enjoy favorable conditions in a mountainous area, especially if the climate is not excessively rainy.

2. *Mediterranean climate.* The entire region along the Mediterranean coast, characterized by olive tree cultivation, is a region all its own. The vegetation here is unique, and features a long period of dearth

---

* For more details, see § 309.

during the summer. In uncultivated areas like woods and scrublands, Mediterranean plants (rosemary, lavender, thyme, savory, etc.) provide an abundant harvest of flavorful, aromatic honey.

3. *Other climates.* It is more difficult to describe the influence of climate in other regions with any precision.

However, all else being equal, we can say that the temperate climates of the west and southwest of France are favorable for stable honey production across successive seasons.

The climate in northern France can be favorable in certain years, but is often too cold or humid; the climate in the east is better.

55. The soil's influence on honey production

The surrounding vegetation, crops, and even climate aren't the only influences on the amount of nectar flowers produce. A given species—buckwheat, for example—will yield varying amounts of honey depending on the soil it grows on.

Generally, then, we can say that, all other conditions being equal, a given plant will not yield the same quantity of honey on all soils.*

Summary

Local nectar resources
Anyone wishing to establish a bee colony should first consider the nectar resources of the surrounding area, within a mile (2-km) radius. If this area is dominated by beet, cereal grain, flax, or hemp crops, or by vineyards, without melliferous trees or meadows, then the area is poorly suited for beekeeping, and one should avoid keeping hives here for any purpose beyond one's own enjoyment.

In other cases, the honey-producing potential of the area depends on the various kinds of plants to be found there, on the climate, and on the nature of the soil.

Influence of climate
Generally, a mountainous climate is favorable for honey production, while the Mediterranean climate makes wintering easier. Among France's other climates, those of the west and southwest offer the most regular continuous honey flow throughout the season.

* For more details, see § 308.

Chapter 5
Setting Up an Apiary

56. The beginning beekeeper and the movable-frame hive

We've already seen the advantages offered by the movable-frame hive, and this is the kind of hive any beginner wishing to pursue modern beekeeping should choose. This hive will allow you to achieve a respectable harvest the most easily, while avoiding the complex operations required to manage fixed-comb hives if they are meant to be productive (§ 195 and following).

But if a beginner can purchase or even build the wooden boxes and frames required,* it's quite rare to acquire moveable-frame hives with the bees already installed. The simplest and best way to learn hive management is to buy some bee stock in fixed-comb hives, and eventually transfer their colonies into movable-frame hives.

This is one initial hurdle that awaits anyone wanting to take up beekeeping. In this chapter and the next, we'll see how to complete the transfer at a leisurely pace, all while making use of the purchased fixed-comb hives to become familiar with handling bees.

We should add that some help from a local experienced beekeeper will greatly shorten the beginner's period of apprenticeship. However, the following primer will assume that the beginner will be left to his own devices, with no one to consult.

---

* See G. de Layens, *Construction économique des ruches à cadres* (ed. Paul Dupont, Paris). (Free plans for building Layens hives and frames are also available from www.HorizontalHive.com. *Ed.*)

57. Stings. Veil and gloves

Almost everyone interested in beekeeping thinks twice when considering the prospect of being riddled with bee stings. This fear is exaggerated. Still, let's consider how we can protect ourselves from stings, avoid being stung at all, and what remedies to use on those rare occasions when we are.

Beginners who must carry out a beekeeping procedure can protect themselves from stings with a wide-brimmed straw hat draped with a black veil whose bottom should be tucked into their clothing. They should also draw their shirtsleeves tight around their wrists using string or a rubber band—and do the same with their pant legs.

A beginner can also wear gloves made of thick canvas or wool—gloves that are sure to be discarded after a certain time, once the keeper becomes accustomed to working with bees. Rubbing one's hands down with a lemon can lessen the chances of being stung.

58. Pacifying bees

To conduct a beekeeping procedure, it's not enough to prevent bee stings; one also needs to avoid aggravating the bees and run the risk of causing others to be stung, even if one avoids being stung oneself.

There's one precaution that is indispensable every time we are preparing to visit a hive: putting the bees into a state in which they are no longer looking to sting. In this state, they beat their wings, generating a loud, intense humming sound, indicating that they have been pacified.

59. Smoking

One can pacify the bees by blowing smoke into their home.

This smoking technique is extremely important for apicultural practice.

If we waft some smoke near a bee by burning a bit of cloth or simply using a cigar, we'll notice the bee agitating its wings to beat free of the smoke—hence, the humming. When smoke is blown inside the hive, the frightened bees gorge themselves on liquid honey, and produce a loud humming sound.

One can smoke a hive by simply lighting a rag and blowing on it—but this is extremely inconvenient. So beekeepers are much better served by using a tool specially designed for this purpose, called a *smoker*.

60. A common smoker

A good smoker is depicted in figure 89.

To use it, we light an old rag, some dry, rotten wood, some brown paper, dry cow dung, or any similar flammable material inserted into the tin cylinder, marked *C*, after raising the cover, marked *CV*. After closing the cover again, we can squeeze the bellows, which one can grip using one hand. The smoke then escapes through the conical tube, marked *T*.

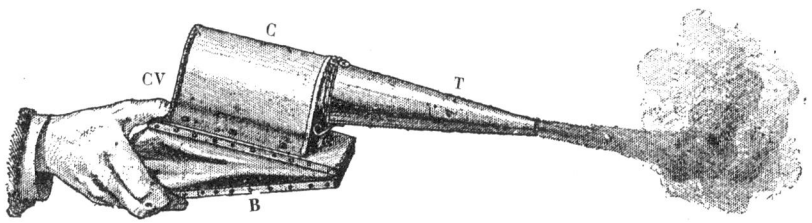

Figure 89. American smoker.

When you aren't using the smoker, you can stand it upright with the tube facing up—but it's a good idea to work the bellows every once in a while to prevent the fire from going out.

A beginner should practice using the smoker before using it with the bees for the first time, to learn how to handle it so that it doesn't go out in the middle of a procedure.

61. Mechanical smoker

The smoker just described is inconvenient in the sense that it only works when the bellows are squeezed, thus occupying one of the operator's hands all too often. Figure 90 shows a *mechanical smoker*, which has the advantage of producing smoke throughout a beekeeping procedure, even without being touched.

This smoker contains a clock movement that can be wound like an ordinary clock. Nowadays these smokers are available for a reasonable price, and can operate for more than twenty minutes—sufficient time for completing a long procedure on a hive. Of course, if more time is needed you can simply rewind the smoker.

The best fuel to use in a mechanical smoker is a strip of cloth—for example, burlap—rolled up and tied up with a string.

This smoker is so convenient that its use is becoming more and more widespread. And since it tends to wear much less than the other models, in the end it can prove a good bargain.

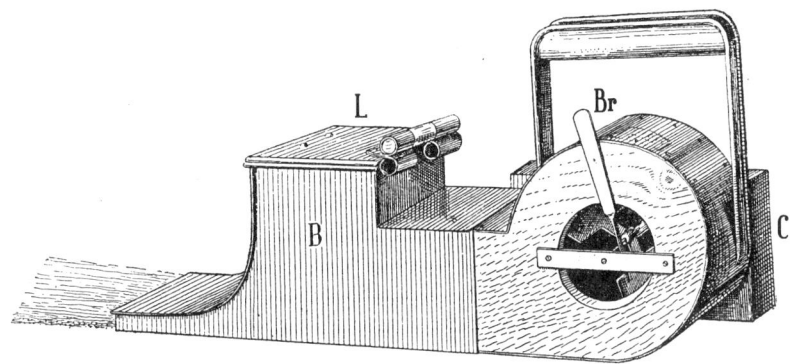

Figure 90. Layens mechanical smoker: *B*, fuel box; *L*, lid; *C*, clock movement; *Br*, brake.

62. How to avoid stings. Sting remedies

You don't have to use a smoker when you're going to check on the bees without opening the hives. In most cases, you can protect yourself from stings by taking the following precautions.

Avoid walking in front of the hive entrance; instead, try to stick to the opposite side.

If you want to observe the bees at the hive entrance, stand still and off to one side, avoiding any abrupt movements; any gesticulations with the arms or head can only agitate the bees.

It's best to inspect hives in the morning or towards the end of the day.

Generally, you can tell that a bee is flying near a visitor with the intent to sting by the higher-pitched sound it makes in flight. In this event, it's best to stoop and move away calmly into some shade, without rushing. After several minutes have passed, the bee will have returned to the hive, and you can return to the bee yard.

If you've been stung, you should move away from the hives unless you're in the middle of a procedure you can't abandon, since the venom from the sting can incite other bees to sting as well.

The first thing to be done is to extract the stinger, suck on the sting, and pinch the surrounding flesh in order to force the venom out. Then, wash the area with some cold water and treat it in one of the following ways.

Cut an onion in half and rub the open face against the sting. You can also crush some leaves of parsley, mint, wormwood, or honeysuckle berries against the sting. Some water mixed with vinegar, water into which you've put some quicklime, some ammonium hydroxide, or, better yet, a drop of carbolic acid, lysol, or thymol, will eliminate the pain.

Or, just after being stung, and having removed the stinger, you can place a lit cigar near the wound until you feel a strong heat; bee venom loses its properties at temperatures of 120°F (50°C) and above.

It can often help to coat the stung area with some honey or oil. If you've been stung a large number of times, then it's a good idea, after removing the stings, to rub the affected areas with alcohol or wrap them with moistened linens.*

63. Precautions to prevent your neighbors from being stung

Generally, the first precaution to take in order to avoid having the neighbors complain about the presence of bees is to maintain good relations with them, giving them the occasional jar of honey or glass of mead.

Beyond this, it's a good idea to arrange the hives carefully so as not to bother the neighbors or passersby.

First, we must point out that if the hives are located in the farmyard, surrounded by trees, large structures, or walls, then ordinarily they will pose no danger for the neighbors, since bees leaving the hive

---

* See also § 223.

for harvest will first have to overcome these obstacles, and won't even think about stinging.

You should also take care not to leave horses or other animals too close to hives that need to be located near a road.

If some animals do happen to be stung, you can rub them vigorously with some straw to remove the stingers, then rub down the affected areas with alcohol or some carbolic acid diluted with water. If these substances aren't immediately available, you can spray the animal with cold water or, if possible, lead them into some water.

64. What irritates bees

When you're inspecting a hive in early spring, you have little to fear from the bees. However, after the main honey flow is over, when the bees can no longer find nectar outside, they can be harder to handle.

Otherwise, during heat waves or during stormy or inclement weather, the bees are more aggressive in the vicinity of their hives.

Of course, an experienced beekeeper will have developed a sense of whether or not a given time is favorable for visiting the hives.

65. Purchasing colonies

Once the location has been chosen and the surrounding area's honey-producing potential assessed, it's time to set up some hives.

For the beginner, the simplest option is to buy them somewhere in the area. But this raises the question: in what condition should hives be purchased, and at what time of the year?

Will any be found for sale in the area? Will sellers be willing to sell in the spring? Can you buy them at some other time besides the harvest? Is it customary in the area to only sell swarms that have just left a hive? Since you may find yourself in any of these situations, we'll now look at them one by one.

In any event, you'll need to know how to assess the value of any hives or swarms for sale, and to learn how to transport them to your chosen location.*

---

* See also § 229.

66. Buying hives in late summer

First, let's assume that hives are available in the area. If you're in a region where *honey merchants* come to purchase fixed-comb hives from beekeepers in order to harvest them and sell their honey, then such purchases will take place near the end of the harvest season. In this case, you'll need to choose this season for buying your hives, since keepers are accustomed to selling then, and a market price is set.*

Even if you're in a region without honey merchants, and hives can be found for sale at any time of the year, harvest time is still the best for buying colonies, since by setting them up appropriately one can make sure that they will winter successfully.

So first we'll assume that the hives will be bought locally, and during the late summer.

67. Assessing the value of hives purchased in late summer

First of all, if possible, beginners should find beekeepers they trust, and who, even at a somewhat higher price, will be willing to sell hives stocked with good, healthy bees.

If the beginner is left to his own devices, how should he go about choosing, or, if no choice is available, appraising the hives' value?

Clearly, a beginner shouldn't proceed exactly as a honey merchant would. A merchant is interested above all in the hives' weight, and is little concerned with the condition of the bees.

The beginner should look to buy hives that are:
1. Well populated with bees.
2. Not entirely filled with honey.
3. Stocked with sufficient reserves for winter.

1. *The purchased hive should be well populated with bees*, since a strong colony will winter better, and have an active population of bees come spring.

---

* Today, bees are usually purchased as nucleus colonies or as packages in spring or early summer. Catching swarms in swarm traps, or bait hives, is another good option for acquiring local stock. *Ed.*

You can tell that a hive contains many bees in the following manner. If you examine a hive during a nice late summer day, the best-populated hives are those with the most foragers leaving or returning to the hive.

2. *The hive should not be entirely filled with honey*, since if the honey reaches too far down in the hive, the bees will be forced to spend the cold season on comb filled with honey, which is bad for wintering.

With the seller's help, you can determine whether a hive is too full of honey in the following manner.

After taking the necessary precautions to avoid being stung, you can begin by lightly smoking the hive through the entrance, then tilting it while continuing to smoke, until you hear a loud humming, indicating that the bees are now pacified (§ 58).

Now, look closely to see whether the combs in the center are empty near the bottom, and at what distance they begin to contain honey, which you can see easily by tilting the comb a bit with your hand (sometimes there may still be some brood on these combs, which is always a good sign).

The hive is in good condition if capped honey (§ 24) only appears in the comb around 6" (15 cm) or more from the bottom of the center combs. At the same time, you can check to see whether the hive's comb is too dark, which, as we know, would indicate that it is too old (§ 30). It is better to buy a hive whose comb is still, for the most part, fairly new.

3. *A hive must be stocked with sufficient reserves for the winter.* You can assess the winter reserves by determining the hive's weight, then subtracting the weight of a similar hive when empty of honey. The difference, representing the weight of the honey in the hive, should never be less than 35 lb (16 kg), and a slightly higher weight would be preferable.

In most cases, if the hive is made of woven wood or straw, the weight of an empty hive will range from around 9 to 13 lb (4–6 kg), so the weight of a purchased hive should be at least 46 lb (21 kg).

We should mention that there are certain regions where fixed-comb hives are very small, and where the desired weight described above will never be found. In this case, you should buy hives that meet the first two conditions—but you may be forced to feed them with sugar in the spring (§ 87).

Also, if you're buying colonies in a large apiary, it will often contain hives of various sizes; all else being equal, you should choose the largest, since they will be best for colony buildup and storing honey (§ 246, II). Moreover, it is good practice to mark the hives one has chosen for purchase in one way or another.

In sum, *don't overlook anything in your effort to acquire good hives*, even if it means paying a bit more; this is the point of departure for your entire operation, and the future of your apiary depends on a strong start.

As the popular saying goes, in order to succeed, you need to begin with "heavy hives, full of flies."

68. Buying hives in late winter

If you find hives for sale in late winter, and if you're able to make sure that they have wintered successfully, then it can be to your advantage to buy them during this season. Any colonies bought on the cusp of spring should meet all of the following conditions:
1. Be well populated with bees.
2. Be stocked with enough honey to last until the good honey flow.
3. Be found to have wintered successfully.

1. You can see that the hives are strong by watching the hive entrances for a while on a nice day when the bees are highly active. The colonies with the highest number of workers leaving and reentering the hive are the strongest.

2. You can judge the honey stocks by weight, as described above. The hive's contents—comb, honey, and bees—should total at least 22 lb (10 kg), after subtracting the weight of the empty hive. This means that a furnished hive should weigh between 29 and 35 lb (13–16 kg).

3. Always taking the necessary steps to avoid being stung, smoke the hive through the entrance and, tilting it, look to see whether the

combs are moldy—if they are, it is preferable to choose another hive. At the same time, with the seller's help, you should spread out the combs and check for compactly arranged brood in the worker cells of the center combs, which is a good sign (§ 137). If the hive only contains drone brood, whether in large cells or in small cells with heavily rounded caps (§ 84), then it is probably queenless, and should not be bought.

When buying hives in spring, you should buy them *more than 1 1/4 mi (2 km) away* from where you plan to locate them, since if hives are bought too close by, a certain number of their bees will return, by habit, to their old location, and be lost for the buyer.

69. When only swarms are available for purchase

There are regions where bee owners refuse to sell their populated hives in both fall and spring, and will only agree to sell swarms (§ 39).

In this case, it is better to buy hives in another location and have them moved (§ 74).

If, however, the only way you can acquire bees is to purchase swarms, then these swarms will be installed directly into movable-frame hives (§ 107). But setting up an apiary by starting with swarms, which almost invariably are bought without sufficient guarantees, is a dangerous game to play, since a swarm often doesn't have enough time to stock up enough honey to survive the winter.*

In any case, you should only buy primary swarms (§ 41), since afterswarms are generally too weak to stock up winter reserves.

70. Prices of hives and swarms

The price of hives varies naturally by region, and from year to year.

A good hive purchased in autumn under the conditions described above will cost between 10 and 20 francs.** Speaking very generally, such hives can be found in Brittany for 10 to 14 francs; in central France for 12 to 16 francs; in Normandy and northern France for

---

* In case of bad weather newly installed swarms (or packages) should be fed— see § 109. *Ed.*

15 to 20 francs; in Champagne and Bourgogne for 16 to 22 francs; around Paris for 20 to 25 francs, etc. Sometimes prices may be higher if the previous year was a poor one. In many areas, hives are sold by the pound, in which case you'll pay between 0.50 and 1 franc per kilogram, after subtracting the weight of the empty hive itself.

The price of swarms can range between 5 and 10 francs. If you're forced to purchase swarms, you should buy the largest ones available, weighing no less than 4 lb (2 kg)—and, as we've said, you should only buy primary swarms.

71. Hive location

Once you've bought some hives as described above, whether in fall or spring, you should be sure everything is ready at their new location before moving them there.

And first, of course, you'll need to choose the best location on your property.

We'll assume that a beginner will start out with three or four hives. This number is sufficient for learning how to handle bees, so a novice shouldn't begin with any more.

If possible, the hives should be situated according to the following conditions:

1. Not too close to one another.
2. Sheltered from the wind.
3. In the shade.
4. Far from a large body of water.

1. During a virgin queen's mating flight, it is of the highest importance that, once fertilized, she not return to the wrong hive—in which case her hive would be orphaned. The virgin queen will be much less likely to confuse the various colonies if they aren't too close

---

** One franc in 1897 could buy about the same amount of consumer goods and services as US$6 would buy today. It could also buy as much gold as $10 would buy today. However, the value of one franc in 1897 *relative to people's income* was much higher, and represented a worker's wage for some 2.75 hrs of work, so purchase of bees was a major investment. *Ed.*

to one another—and the workers will also be able to locate their hive more easily when they return from their usual foraging flights.

So contrary to what one often sees, it's a good practice to set up your hives as far away from one another as possible, by several yards (meters)—and if that's not possible, you should avoid arranging them in anything resembling straight rows.

2. As we've seen, bees are tired when they return from foraging—and as they approach the hive, the wind can sometimes drive the exhausted bees to the ground. If the weather is cold they may never become airborne again. So it's best to make sure the hives are protected from the prevailing winds by a building or wall, or by some trees.

3. During hot spells, the wax in honeycomb can sometimes grow soft, and the combs can sag.

So, when possible, it's better to set up your colonies in the shade, not in direct sunlight.

Colonies do well when shaded by trees, and even, when possible, in a forest, which in fact is their natural habitat, provided that they're near the forest's edge.

4. If possible, you should avoid positioning the hives in the direct vicinity of a large river or lake, since a broad expanse of water can interfere with the bee's flight, and wind can cause them to drown there.

72. Hive stand. Bottom board

To avoid humidity, hives should be elevated at a certain height off the ground, so you should set up *hive stands* before installing the hives.

Since we've assumed that the beginner will be installing on these stands fixed-comb hives that will later be turned into movable-frame hives, it's best to cover the stands with *bottom boards* that will be suitable for holding frame hives from the start. In fact, this will prove beneficial when the transfer occurs, since the bees will easily recognize their old bottom board; when the common hive is replaced by a frame hive, they'll more easily find their way back to their new home since the bottom board will be the same, and familiar to the bees. The props can be made of bricks or rocks, or, even better, of wood.

Figure 91 shows a wooden stand, *S*, with a bottom board, *B*; it has the advantage of being easy to move.

The stand should be positioned so as to allow you to move freely around the hive.

Figure 91. A stand, *S*, with a bottom board, *B*.

73. The water basin

We've seen (§ 19) how bees need to gather water in order to thin down their honey and prepare food for their larvae. If there's no water to be found near your hive, in streams, ditches or ponds, then it's desirable to set up a *water basin* for the bees.

The basin can consist, for example, of a tub or the bottom of a barrel in which water is kept, along with some floating bits of wood or corks on which the bees can alight when they come to gather water.

74. Transporting hives

Now that the hives have been purchased, and the stands with bottom boards readied for them, your task is to transport them to their new location.

If the hives are bought in the fall, not far from their final location, you can simply move them on a wheelbarrow, in the following way.

First, wait with moving the hives until the bees are no longer venturing forth from their hive on a daily basis—although you shouldn't wait until frosts begin since any bees that separate from the cluster during the move will be unable to rejoin it due to the cold.

When you're ready to transport the hives, take some burlap large enough to wrap them completely.

As evening approaches on the day before the move, after lightly smoking each hive through the entrance, lift it and spread the burlap across the bottom board, then lower the hive back onto the canvas.

Hives should be wrapped for transport after sunset, in order to be certain that all of the bees have returned to the hive.

If necessary, gently smoke the hive once again, then wrap it completely using the burlap. Fold and tie it up so as to make sure that no bee can escape, while still leaving them room to breathe thanks to the air circulating through the loose mesh of the cloth.

Then, place a wedge beneath one end of the wrapped hive's base (fig. 92) to allow air to circulate.

When each hive to be transported has been readied in this way, line the wheelbarrow with some straw. Then, place the hive on the bed of straw, packing some straw on each side, such that the combs are positioned vertically and in the direction of the wheelbarrow—this helps prevent the individual combs from bruising against each other during the move. Also, put a couple of sticks of wood beneath the hive to ensure the circulation of fresh air. During transport, avoid any jolts or abrupt movements that might excessively shake up the bees.

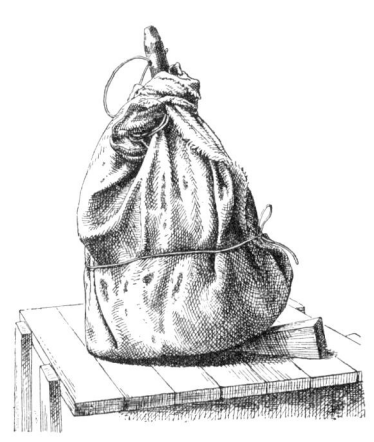

Figure 92. A hive wrapped and ready for transport.

Once you've reached the stand where the hive is to be installed, set it still wrapped atop the bottom board, still propped up on one side by a wedge; smoke the hive a bit through the canvas, before unwrapping it cautiously. Finally, put the straw hackle on the hive.

If you find bees clinging to the canvas, detach them gently using a goose feather, for example, causing them to tumble down onto the bottom board, where they'll be able to rejoin the others; then, remove the wedge.

Now your hive has been successfully moved and installed.

CHAPTER 5. SETTING UP AN APIARY

Of course, for all such procedures—spreading the canvas, wrapping, and installation—a beginner should be outfitted with a veiled hat.

A wrapped hive can also be moved on foot (fig. 93), carried across one's shoulder on the end of a stick, or in a basket. In the case of the latter, the wrapped hive should be situated so that it is upside down in the basket.

If the hives are being moved a greater distance, or if you're going to the train station to pick up hives purchased from further away, then you'll need to transport them by wagon (or by mule in mountainous regions without accessible roads).

Figure 93. Hive transport on foot.

The wagon you use should have good suspension. Line it with a thick bed of straw, and place some sticks on the straw to support the wrapped hives as described above; keep them tightly arranged, right up alongside each other, and tied down to the straw with ropes to avoid being shaken en route.

If you're moving them a great distance, then you can buy some commercially available canvas with a wire netting in the middle to allow the bees to breathe more easily.

If it's hot, it can be very dangerous to move the bees during the day—in any case, it would be a good idea to wait until nightfall.

If the purchased hive contains recently built and therefore fragile comb, you should take great precaution during transport to avoid any jolts that could break the combs.

75. Transporting swarms

If you've had to purchase swarms, and are moving them a short distance, you can do so quite simply, in the evening, in a carefully wrapped skep in which the swarm has been gathered.

If you've bought swarms at a distant location, you should provide the seller in advance with some cases built specifically for this purpose.

The bottom of a transport case should be replaced by a wire mesh with rather small gaps that prevent the bees from getting through, and the case should be covered with a top that also has such a wire mesh.

The seller should gather the swarm in this case just as he would in a skep; once all the bees are inside, he should close the case and deliver it.

As we've said, if you have no choice but to purchase a swarm, it should be installed directly in a frame hive (§ 107).

76. Wintering purchased fixed-comb hives

Let's say that you've bought some colonies in autumn. Once they've been transported and installed, you'll need to prepare them for wintering.

Make no mistake: *successful wintering is critical to successful beekeeping.*

All too often, beekeepers don't know how to winter their hives, and this is a source of many disappointments in beekeeping.

Since wintering is so critical, a beginner can't devote too much attention to this first procedure, on which the very future of the newly purchased hives depends.

The first thing to understand about wintering is that the bees clustered in their hive fear humidity more than the cold; all things considered, it would be better for a hive to be exposed to interior air currents during the winter than hermetically sealed and draft-proof. If your biggest worry is the cold, you might seal off the hive on all sides, save for an extremely small entrance, such that the air in the hive cannot easily be refreshed. In this case, you run the risk, come spring, of finding the moldy comb and sick bees, and often many dead ones—an unsuccessful wintering indeed.

Here's how you should go about wintering—in both the best and the simplest way.

In the evening, having removed the hive's hackle, gently lift the hive and place underneath it three wedges, around 3/16" (5 mm)

CHAPTER 5. SETTING UP AN APIARY

thick—for example, bits of slate—one to the left, one to the right, and one in the back. This will allow the air to circulate perfectly in the hive throughout the winter season.

But the front of the hive, and its entry, could provide an opening for field mice to enter during the winter. To prevent them from penetrating the hive, cut out a strip of perforated metal or wire netting (fig. 94) that will allow the bees to exit, but keep their enemies from entering. Attach this metal strip with wire, such that it is flush with the bottom board and covers the lower portion of the hive. By the way, instead of perforated metal, you could use a series of long, thin nails, hammered into the bottom board in front of the entrance.

Figure 94. A skep during wintering.

Now, put the hackle back on, which, if possible, should extend lower than the bottom board; then, press the hackle's straw against the hive using a barrel ring, for example.

Outfitted in this fashion, the hive is ready for wintering: it has nothing to fear from a lack of fresh air, or humidity, or mice, and the straw hackle will protect the hive from the cold and the rain.

Having taken these precautions, you can leave the hives untouched throughout the winter.

In areas where snowstorms are possible, it's better not to winterize as described, since fine snow can make its way into the hive and accumulate there.

You can leave the hive without the wedges underneath, simply placing a strip of perforated metal in front of the entrance as just described; and providing another means of supplying the hive with fresh air. The simplest way is to replace the ordinary bottom board with one that has a square hole cut into it, 6" (15 cm) on each side, covered with wire netting.

Summary

Precautions against stinging

To carry out initial procedures, beginners should equip themselves with a veil and gloves to protect against stings, and learn to handle a smoker, which will help control the bees.

Buying hives

When you're just getting started with beekeeping, it's wise to begin with a small number of hives. Once you've chosen a location for them, protected from wind and in the shade if possible, it's best to buy already populated hives rather than swarms. Preferably, you should make this purchase in the late summer.

A hive bought during this period should be strong and not entirely full of honey—yet with enough honey reserves to last the winter.

Before installing the hives, set up some stands with bottom boards at the desired location. Then, transport the purchased hives, after wrapping them, and having taken all the necessary precautions.

Wintering

Once the hives have been transported and installed, prepare them for successful wintering—a procedure that is absolutely critical for successful beekeeping. The hives should be set up so as not to fear a lack of air, or humidity, or predators. Their straw hackle protects them against the rain and cold, and you can leave them to wait out the winter without touching them again.

Chapter 6

What to Do in the Spring Of Year One

77. A beginner's apprenticeship

As a beginner, the most important thing is simply to become comfortable working with bees. During year one, you'll have to dedicate a fair amount of time to carrying out various procedures, and you'll need to inspect your hives at various points in the year. In short, you'll need to get used to handling bees—a habit that is essential for any beekeeper. Think of it as your apprenticeship, and don't spare your time or energy in developing the skills you'll need to manage your bees in the future with confidence, and with minimal time and energy.

It's worth remembering that no serious knowledge can be gained without effort. Likewise, in this branch of agriculture as in any other, establishing a reliable source of revenue takes some work.

Beginners who plan to establish an apiary using movable-frame hives are almost always forced to start out with fixed-comb hives. Since beginners often find themselves dealing with fixed-comb hives even when their bees will eventually be installed in movable-frame hives, it's a good idea to embark on your beekeeping apprenticeship using fixed-comb hives. You might say that learning to operate fixed-comb hives is the first step to becoming a skilled keeper of movable-frame ones.

78. The end of the wintering for hives bought the previous autumn

When the first flowers start to bloom after winter has passed—that is, when the willows, poplars, apricot trees, violets, wallflowers

(*Erysimum cheiri*), or windflowers (*Anemone*) begin to blossom—then the time has come to de-winterize and inspect the hives you bought the previous autumn.

You can begin by removing the wedges you placed between the hive and the bottom board, as well as the mouse guards (strips of perforated metal or wire netting) you attached with wires. If your hive was placed atop a special winter bottom board with a hole covered with wire netting, you can replace this winter bottom board with an ordinary one.

The gap we left between the hive and the bottom board to encourage air circulation during the winter is no longer needed, now that the bees will be venturing from the hive on a daily basis; in fact, the hive interior will need to be very warm in order to encourage brood development.

You should wait to inspect the colonies until you've seen the bees actively exiting the hive to collect nectar, pollen or water for a week or so, in order to give them some time to reorganize for the new season that is now beginning.*

Let's assume that we'll visit the hives on a nice day, when the bees are highly active, and inspect each hive, one by one. This springtime inspection is indispensable, since we need an exact knowledge of the condition each hive is in before we carry out the various procedures that lie ahead.

Now let's consider all of the special cases that may arise during this inspection.

79. A hive in excellent condition after wintering. Inspecting a fixed-comb hive**

With the hive placed on a stand, let's smoke the bees around the hive entrance (fig. 97), then prop up the hive on a wedge an inch or more

---

* During inspections that are performed too early, before the hive is back on a work footing, bees have on occasion been observed to kill their queen.

** For this inspection, you'll need the following: 1) a hat with a veil and a smoker that is ready for use; 2) a long kitchen knife or, better yet, the straight and hooked knives shown in figures 95 and 96; 3) a stool or stepladder; 4) a goose feather or a bee brush (fig. 129), a gimlet or drill and a long piece of stiff wire; 5) a notebook and pencil.

(several centimeters) thick and continue smoking lightly until you hear a loud hum (§ 58). Now, turn the hive upside down, holding it steady between the legs of an upside-down stool. Scrape the bottom board clean of any debris or dead bees you find there. Don't forget to blow a bit of smoke into the hive from time to time to keep the bees pacified.

As you drive back the cluster of bees with smoke (a skill the beginner can easily master), the bare combs will become visible, allowing you to examine them carefully. First, we should make sure that the wax isn't moldy. As we continue to hold the bees at bay with some

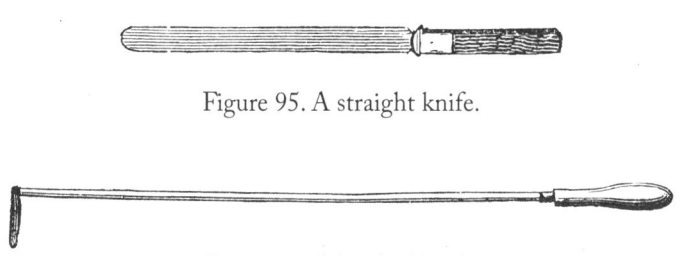

Figure 95. A straight knife.

Figure 96. A hooked knife.

Figure 97. Smoking a fixed-comb hive.

smoke, let's turn our attention to the middle of the center combs to see whether there is any sealed brood there (*c*, fig. 36). If we don't see any, then we shouldn't hesitate to make a deep cut into one of these center combs using a long kitchen knife, in order to remove a piece (fig. 98); then we can use a goose feather to sweep any bees that may be clinging to this piece of comb down into the upside-down hive.

As we look over this piece of comb, we can see some sealed worker brood (§ 26), or at least some eggs and larvae at multiple stages of development in the worker cells. This proves that the hive has a queen,

Figure 98. Inspecting a fixed-comb hive.

and the cluster of bees, filling four or five gaps between combs, shows us that the population is strong. Let's put the hive back on its stand. As for the honey reserves remaining in the hive, we've already deduced from the weight that they are likely sufficient; if we want to make sure, all we have to do is use a gimlet to poke a hole in the upper third of the hive, then insert a thick wire through it; when we pull it out, it should be coated with honey.

If we return the next morning to examine the hive that has been set up in this fashion, and the weather is still as nice as the day before, we'll see a large number of workers leaving and returning to the hive. Many of them are bringing back pollen.

If the results of your visit are like those described above, then your colony is in excellent shape. Take note of this fact in a notebook, under the identification number given to this hive.

80. A weak hive that has wintered well

When you inspect a hive as we've just done, you may find that it isn't as strong as the first one described. Instead, you find a dense cluster of bees near the center of the hive; there is also some worker brood there. Make sure, as we did before, that the comb isn't moldy, and that the honey reserves are sufficient. When you watch the bees at the hive entrance on a nice spring day, you see a coming-and-going of bees that, while fewer in number, have the same vigorous energy level as that of the stronger hive.

This colony is weak, but it has wintered well. If it has a good queen, its population may well grow considerably during the foraging season, until it catches up with the strongest colonies.

81. A strong hive that has wintered poorly

If while inspecting a hive you find many dead bees that have fallen onto the bottom board, and others that have accumulated between the combs and are blocking the circulation of air—and, in addition, you find many moldy combs—then you can conclude that the given hive has wintered poorly. Without a doubt, this means you didn't take the ventilation-related precautions detailed above, or that, for one reason or another, the gap left beneath the hive became obstructed.

Nevertheless, you find a considerable number of bees in the hive, and confirm that worker brood is present.

Let's remove the moldy comb by cutting it out, and use a goose feather to sweep off all of the dead bees who are blocking the gaps between the combs; finally, scrape the bottom board free of any debris and reinstall the hive just as you did the others. This colony is still strong, but judging by the large numbers of dead bees and the humidity inside, its population has dwindled during the winter, and many of the workers are likely still sickly. The colony may bounce back; but it's also possible that this once-strong hive will remain mediocre.

82. A hive that has run out of honey

You may have another, more sinister situation on your hands: a hive from which no bees can be seen emerging. As we remove the mouse guards and the winter wedges and lift the hive—which weighs very little—we find many bees, apparently dead, piled atop the bottom board. The bees found in between the comb are also motionless, and many of them are head-first inside empty cells. This is a hive that has run out of honey and starved.

Are the bees dead, or completely stunned? When we attempt to warm some of them with our breath, we notice some movement, and suddenly we hope that the colony may still be saved.

To this end, we sweep the bees from the bottom board back into the hive, since they may still be alive. Then, we carefully wrap the hive—still upside-down—with some burlap and move it into a warm room.

Now prepare warm sugar syrup—half sugar, half water—and pour a glass of it onto the surface of the canvas, atop the upside-down hive. If most of the bees are merely stunned, they'll be reanimated by the room's warmth and the sugar syrup fed to them through the cloth. When evening falls, move the hive back to the apiary, turn it right-side-up and set it atop its stand without removing the canvas, and inserting one of the wedges on one side in order to allow air to circulate. The next morning, lightly smoke the hive before removing the wrapping and the wedge, and continue to feed it for several more days (see § 87, etc.).

83. A dead hive

If, in a situation like the one just described, the bees cannot be revived, then the colony has starved to death. But you may even find a dead colony in a hive that still contains a lot of honey. It may be that the colony became queenless when winter set in, or for whatever unfortunate reason. In this case, you'll find the dead bees crowded on combs that are completely empty of honey—and yet, very close by, you may find some combs full of capped honey near the hive's walls. How can we explain such a seemingly strange situation?

Quite simply, there was no honey stored away near the top of the combs with the bees—honey they could have fed on as they made their way upward during a prolonged period of cold, when there wasn't a single day warm enough to allow them to cross over to other combs, still full of honey, in another section of the hive. Once they had consumed all of the honey to be found where they were, they died of starvation, unable to reach the remaining stockpiles. As we'll see later, this is one reason why it's advisable to have hives whose combs are relatively large and taller than they are wide.

Later, we'll discuss what to do with this dead hive (§ 85).

84. A hive that is queenless or has a drone-laying queen

You may find that one hive seems far less active than the others—and the few bees you see exiting and entering don't seem particularly anxious or hurried. From time to time, you notice a worker who, instead of busily leaving the hive and setting out boldly in a certain direction, seems unsure of where to go. In the same way, a returning bee seems to hesitate instead of rapidly crossing the entrance. Sometimes, you may notice something that is very strange indeed: a bee *leaving* the hive with pollen. You never notice the bees performing the "play flight" (§ 11). Still, the hive contains honey reserves. What might explain this lack of activity?

Let's smoke the hive and take a look inside. Even when we search near the top of the combs, we don't find any brood—or, if we do, it's exclusively drone brood (§ 27), either in drone cells or even in worker cells, in which case the cap will bulge further outward (*D*, fig. 52).

All of these signs tell us that the hive is either queenless or has a drone-laying queen (§ 35).

However, it may still be the case that the eggs or larvae in the small cells might be worker eggs or larvae—we can't know for sure right now. Out of caution, let's simply take note of the hive's condition and inspect it again two weeks or so later. If at that point we don't find any sealed worker brood, then we know for certain that the hive is definitely doomed—that is, it has a drone-laying queen or no queen at all.

85. What should be done with a dead or disorganized hive?

A disorganized hive that is left standing in an apiary has no hope of recovering—either it has no queen, or it has one that is incapable of laying worker eggs. Should you simply leave it where it is? No, since it could be robbed—not to mention the fact that the bees still in the hive could be of service in the other hives. And nothing could be easier than winning them acceptance to the remaining colonies.

Pick a nice day when the bees are highly active; lift the disorganized hive after smoking it lightly, then knock it against the ground to shake out the bees, who, unable to relocate their former home, will find their way into neighboring hives.

The bee-less hive—the dead hive we discussed earlier—should be moved indoors until you can make use of its contents.*

86. Treating combs with sulfur

When a hive of bees has been removed from an apiary as just discussed, you'll need to burn a sulfur wick inside it in order to eradicate any eggs of wax moths (§ 290) that might later develop and destroy the comb.** Here's how to do it.

Dig a hole in the ground that is a bit smaller than the hive is wide, and about 6" (15 cm) deep. Then, attach a bit of sulfur wick to the end of a wire, stick the wire into the ground in the middle of the hole, light the sulfur, and cover the entire hole with the hive, shoveling some dirt around the sides to seal it. The procedure will be complete

---

* *Note:* The springtime inspection as described here for fixed-comb hives is much simpler with moveable-frame hives. Indeed, nothing is easier than inspecting the brood, assessing honey levels, and determining overall colony condition with moveable-frame hives. The possible conditions of moveable-frame hives during the spring inspection are the same as those we've just described for fixed-comb hives.

** Paradichlorobenzene (active ingredient in modern mothballs) is the chemical currently used in the U.S. for preventing wax moth damage to stored comb. Combs need to be thoroughly aired before returning them to the bees. Freezing empty comb and then storing it in a cold well-ventilated space is a good chemical-free alternative. *Ed.*

after half an hour. Pick up the hive, put it in a well-enclosed place, and disassemble the combs.

To remove the combs, cut them loose one by one with a hooked knife, setting aside those that contain honey. Combs containing drone cells and brood should be placed in boiling water to form into wax balls that can be used later to melt (§ 277). Now all that's left are empty combs consisting of worker cells that can be used to prime frames (§ 100).

87. Feeding hives that are low on honey

If your hives were purchased strictly according to the conditions previously described, they will have enough honey to last until the warm season, and you'll have nothing to worry about in this regard. You'll only need to come to the aid of a hive during the spring by feeding it with sugar syrup if it was bought in the fall and was too small or lacked sufficient honey reserves.

88. Which hives require feeding?

First you need to determine which hives need to be fed. You can do this in two ways: 1) by the hive's weight; 2) by probing it with a wire.

1. *By the hive's weight.* Weigh the hive, then subtract the weight of a similar hive when empty—a weight you should have determined when you first bought the hive. Then, for a hive of approximately 1,800 cubic inches (8 gal or 30 L),* subtract 3 lb (1.5 kg, the weight of the wax) and another 3 lb (1.5 kg, the weight of the bees and the brood). For normal-sized fixed-comb hives, if the remaining amount is less than 11 lb (5 kg, the weight of the honey), then it would be wise to feed this hive before foraging season begins.

For example: We know that a hive weighs approximately 9 lb (4 kg) when completely empty, and find that its weight is now 22 lb (10 kg), including all of its contents. So we need to subtract 9 lb (4 kg, the

---

* If the hive has a larger capacity, subtract a proportionately larger weight to account for the weight of the wax and that of the bees and brood.

weight of the empty hive), 3 lb (1.5 kg, the weight of the wax), 3 lb (1.5 kg, the weight of the bees and the brood) from the total of 22 lb (10 kg), leaving 7 lb (3 kg, the weight of the honey). Since this is less than 11 lb (5 kg), we know it won't be enough to last until the honey flow: we'll need to feed the hive.

2. *By probing.* On occasion, you can also probe a hive using a wire (§ 70). When the probe indicates that the only remaining honey is near the very top of the hive, then you'll need to feed the hive. This method is easier than the previous one.

89. Feeding a fixed-comb hive

The simplest way to feed a fixed-comb hive is as follows.

First, make some syrup—half sugar, half water. If we heat the solution, the sugar will melt more quickly. Take a bowl that is small enough to be covered by the hive. In the evening, smoke the hive that needs to be fed, tilt it to one side, and try to place the bowl underneath the

Figure 99. Feeding a fixed-comb hive.

combs (fig. 99). Generally, one of two things will happen when you lower the hive back into position: either the hive walls won't rest flush atop the bottom board because the combs reach too far down, causing the bowl to prop up the hive a bit above its stand; or the entire bottom of the hive will rest flush against the stand, but the bottom of the combs won't reach the rim of the bowl.

1. If the combs reach all the way to the bottom of the hive (fig. 100), set the hive atop the bowl.
2. If the combs don't reach low enough, prop the bowl up on some bits of wood so that the bottoms of the combs touch its rim.

Now, pour 1 lb (0.5 kg) of warm syrup in the bowl—or even 2 lb (1 kg) if the hive is large. Cut up some slices of cork and put them on top of the syrup (a single cork can make four or five such slices)—or, if you don't have cork, many bits of straw—to allow the bees to feed easily without getting stuck in the syrup.

You should perform this procedure around nightfall, when all the bees have just returned to the hive, in order to avoid robbing—that is, an attack on the hive you're feeding by bees from other colonies (see § 92 below).

Figure 100. A fixed-comb hive completely filled with comb, viewed from below.

Indeed, robbing is the greatest danger when it comes to feeding, and when it happens it can lead to battles between all the hives, which can be very discouraging for a beginner. So we can't stress enough how important it is to take every precaution against possible robbing—in fact, we suggested buying hives well stocked with honey precisely in order to avoid having to feed, and all of the resulting worries, especially the danger of robbing.

Preventing robbing is another reason why, quite early the next morning, before the bees have left the hive, it's critical to remove the bowls you've set underneath the hives the night before. *You must remove the bowls even though they may still contain syrup and bees.*

Put the bowl in a room, and any bees remaining there will eventually fly out through the window and rejoin their hive.

90. How bees feed on syrup

When the bowl of syrup is placed near a group of bees, they'll usually hurry onto the bits of cork, and the colony will quickly absorb the food.

Meanwhile, the bees will imagine that there's nectar to be found in the flowers, and several will leave the hive to go foraging, but will return when they see that it's nighttime outside, and will resume feeding on the syrup along with the others.

However, it may happen that the bees won't touch the sugary liquid in the bowl.

If it's too cold, or if the hive is too weak, the bees clustered near the top of the hive won't crawl down to take the syrup, as you'll be able to tell when you lift up the hive the next morning to remove the bowl. In this case, go ahead and remove the bowl, but when you put it back in the evening, turn the hive over and drizzle several spoonfuls of the syrup in between the combs. This is the best way to lure the bees down toward the syrup.

If the hives you've bought have an opening on top—as with cap hives for example—then they can be fed more easily from the top, through the opening, in the following way.

Put the syrup in a mason jar with the top covered with a fine-mesh cloth, expose the opening at the top of the hive, and place the jar upside-down against the opening before recovering the hive with its hackle. The next morning, remove the jar just as you would remove the bowl.*

91. When should you stop feeding?

In this way, give the bees between 1 lb and 1 3/4 lb (500–800 g) of syrup per week, depending on hive strength.

Continue feeding in this way until the first major honey flow—that is, until swarming season begins.

92. Robbing

As we've just seen, a hive that is being fed syrup always runs the risk of being robbed. If you fail to take the precautions we've just recommended—if, for example, you forget to remove the syrup bowl early enough in the morning—then the hives you're feeding may well be robbed. The bees from other colonies will notice the workers actively leaving this particular hive as if in search of honey, and will surmise that these bees are gathering nectar while they themselves aren't.

If a bee from another hive manages to get into the hive that's being fed, she will load up with the sugar syrup before going to pass the word to the other bees in her own colony. You'll notice the bees of this colony arriving at the hive that is being fed in greater numbers, and the bees guarding it will begin to worry.

The besieged bees will gather near the hive entrance, and the battle will begin. The workers will engage in hand-to-hand combat, trying to sting each other.

If the fighting is allowed to continue, regardless of which side is winning—the robbers or those being robbed—then the apiary may face grave consequences. This battle may excite the bees in the remaining colonies and lead to widespread fighting within the apiary.

---

* The feeding procedure we've just described for fixed-comb hives is simplified considerably for moveable-frame hives. In fact, all you have to do is give weak hives some honey frames taken from hives with too many.

Robbing doesn't only threaten when you're feeding a hive. Bees may also try to invade hives that are queenless or extremely weak. Any honey left within the bees' reach or in a room with an open door or window may provoke robbing. Finally, if a beekeeping procedure like a hive inspection drags on for too long, then there will be a risk of robbing.

As we've seen, robbing with fighting is the result of an oversight or an imperfectly taken precaution; as we continue to master the art of practical apiculture, we'll see that *it is always within a beekeeper's power to prevent robbing.*

93. How to stop robbing

Such fighting among bees must be stopped at all costs. The first step is to narrow the entrances of all the hives, such that no more than two bees can enter the hive side by side at a time. This helps to prevent robbing bees from getting into the other hives.

As for a robbed hive, the simplest and most reliable step is to smoke it, wrap it in canvas, and move it to a cellar, propping it up on a wooden wedge to be sure it gets enough air. Leave it there in the cellar, and return it to its former spot the next evening once all the bees have returned to their hives. Keep the entrance narrowed for several more days, and don't resume feeding until the apiary has calmed down.

Sometimes robbing can be stopped without removing the hive, in the following manner: after narrowing all the entrances, spray the bees in the agitated hives with water, then spread some kerosene on the surface of the robbed hive and on its stand.

94. Pollen substitutes

As we've just seen, hives can run short of honey in the spring. Sometimes, in certain regions, they may not be able to find the pollen they need to feed their young. In this case, you'll have to supplement the hive as follows.

Put some flour (a good pollen substitute) within the bees' reach; they prefer rye flour. Keep it in some small boxes to protect it from the wind. To keep the bees from drowning in the flour, nail some strips of wood to the bottom of the boxes and pour the flour in the remaining gaps. Take care to remove the boxes each evening to keep the humidity from causing the flour to lump.

It must be said that this procedure isn't always completely necessary, and may cause complications if the flour is rancid or of poor quality.

Summary

A beginner's task
The first job of any beginner is to become comfortable handling bees. During the first year, beginners should gain experience with a large number of the various procedures involved in managing their colonies. Beginners can only begin to simplify their management of the hives and save as much time as possible once they've gained the necessary level of experience.

Springtime hive inspections
The first thing to do in spring is to inspect the hives after wintering. Using a few simple tools, the beginner will learn to assess—and record in a notebook—the condition of the beeyard after winter, assigning each hive to one of the following categories: hives in excellent condition; hives that are weak but that have wintered successfully; strong hives that have wintered poorly; hives that have run out of honey; and hives that are dead or disorganized. Hives in the latter two categories should have their combs treated with sulfur, and any comb or honey that can be put to use should be placed in a well-sealed area.

Feeding fixed-comb hives
If your hives weren't purchased the previous autumn in accordance with the conditions we recommended, they may run low on honey in the spring, and you'll have to feed them.

You can feed fixed-comb hives by correctly placing a bowl of syrup beneath the hive at nightfall and removing it early the next morning.

Robbing
If you fail to remove the syrup as recommended, you'll run the risk of robbing. This is one of the biggest problems faced by beginners—but keep in mind that it is always caused by negligence.

If robbing has occurred, it must be stopped immediately. The entrances to all the hives must be narrowed; the robbed hive should be smoked, wrapped, and moved to a cellar, then put back in place the next morning with its entrance narrowed until things are back to normal in the apiary.

Chapter 7

Installing Swarms In Movable-Frame Hives

95. The honey flow

It's impossible to predict the exact timing of the honey flow—it depends on the weather and the kinds of honey plants found in the region.

Generally speaking, if many of the local honey plants are in bloom and there is a string of nice warm days after a rainy period, then the bees will gather a large amount of honey.

But you need only take a look at the hives in order to see whether the honey flow is underway. There will be many more bees exiting and entering the hives, and you'll see a lot of workers landing clumsily on the landing board in front of the hive before entering—which, as we know, indicates that they're loaded down with honey.

96. Various methods for judging how the honey flow is going

It's interesting to track the fluctuations in honey flow—and you can do so in several ways:
1. By the bees' overall activity level.
2. By the number of fanning bees.
3. By the number of bees gathering water.
4. By the weight of the hive.

1. *By the bees' overall activity level.* By carefully watching the bees leaving the hive, and taking note, for example, of the number of bees

loaded down with honey that return per minute, you can get a rough idea of how much honey is being gathered at various hours of the day. On a day with high honey flow, you'll find the bees to be highly active early in the morning, somewhat fewer in number around noon, and quite active again in the afternoon until nightfall.

2. *By the number of fanning bees.* We've already seen (§ 6) the ventilating bees at the hive entrance that beat their wings to create air circulation inside the hive after a major harvest. Ventilating bees are only present when the bees have just gathered honey; the air flow they create is meant to cause the excess liquid in the nectar, freshly deposited in the cells, to evaporate. The more honey has just been harvested, the stronger the air current must be, all else being equal. So if you count the number of fanning bees, at the same time of day—in the evening when the bees have returned to the hive, or in the morning before they leave—you'll get an idea of how the harvest is going.

The number of fanning bees can also help you determine which hives are bringing in the most honey.*

3. *By the number of bees gathering water.* If you've installed a water trough for your bees (§ 73), you can also get a sense of how the foraging is going by the number of bees gathering water. If almost no honey is being collected, there will be many bees at the trough, but if a lot of honey is being collected, you won't see any more bees there. This is easily explained if we recall that freshly collected honey always contains excess water. Since this excess effectively replaces the water the bees had to gather outside when no honey was being gathered, you won't see any more bees at the trough during a strong honey flow.

4. *By the weight of the hive.* If the hive is installed atop a scale (§ 219), you can also judge how much honey is being collected during intense honey flow by the hive's weight, in the evening, once all the bees have returned.**

---

* See G. de Layens, *Étude sur la ventilation des abeilles* (in *Apiculteur*, January 1896).

97. Preparing movable-frame hives for installing swarms

When foraging season begins, you should begin to prepare for any swarms that might emerge from your hives. A beginner should collect these swarms for installation in movable-frame hives.***

So you need to: 1) prepare movable-frame hives to receive the swarms; and 2) be ready to collect the swarms when they emerge from the hives.

98. Description of a movable-frame hive

We've already described (§ 46) the general makeup of a movable-frame hive, but now the time has come to handle it on a practical level. So we need to select a particular model and learn how it works, down to every last detail.

There are many movable-frame hive systems (§ 211). The hive we're about to describe is one of the best for simple and productive beekeeping.****

This hive (fig. 101) consists of a bottomless wooden case whose top, which serves as the hive's roof (T, fig. 101), is attached to the case by two hinges as seen in the illustration. The two largest faces of the case

---

** You shouldn't assume that an increase in weight during the day will correspond to an increase in weight of capped honey; in fact, this rise in weight is due to the newly collected honey, from which the excess water is being evaporated due to the air currents produced by the ventilating bees. This explains why, during a heavy honey flow, you may find that a hive has lost significant weight overnight—it will weigh less the next morning than it did the previous evening.

*** Unless, of course, the colonies were transferred in spring—that is, moved into a new hive by: direct transfer (§ 144), superposition transfer (§ 230, point 1), flipped-hive transfer (§ 143), or by artificial swarming (§ 230, point 2). The simplest transfer method is the flipped-hive transfer, but it usually only succeeds with strong hives, and during a heavy honey flow. The quickest transfer procedure is the direct transfer, but it can prove rather difficult for a beginner.

**** This hive, of the horizontal type that is sometimes referred to as a *French hive*, is known in retail as a *Layens hive*—a less apt designation than the first. (The Layens hive is defined by the following key dimensions: interior frame length 31 cm (12 3/16"), interior frame depth 37 cm (14 9/16"), and comb spacing 38 mm (1 1/2") on center. *Ed.*)

CHAPTER 7. MOVABLE-FRAME HIVES. SWARMS

are referred to as the *front* and *back* of the hive, while the two smallest faces are called the *sides*, and the case in its entirety constitutes the hive *body* (*B, B*, fig. 101). Figure 103 depicts the hive body alone. A major portion of the hive's front and back is covered with straw, as seen in figure 101. The hive's body contains the wooden *frames* as seen in figure 104. These frames, twenty in number, are positioned parallel to the hive's sides. One such frame, marked *F* in figure 103, can be seen in position inside the hive body.

Finally, this bottomless case simply rests on a board that juts out in front and is called the hive's *bottom board* (*bb*, fig. 101); another small board, *f*, affixed to the front left-hand side of the bottom board, serves as a kind of landing pad for the bees.

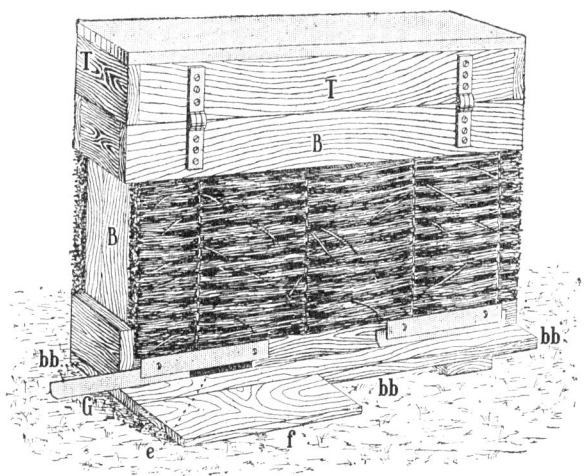

Figure 101. A horizontal moveable-frame hive with flat roof and hinges. *T*, roof; *B*, hive body; *bb*, bottom board; *f*, flight board; *e*, entrance; *G*, metal tab for narrowing or widening the entrance.

The flat roof on hinges can be replaced by a pitched roof, either with hinges (see above fig. 66, § 46) or without (fig. 101).

Now, let's describe each of the hive's major parts.

The front and back of the hive each consists of a board with a protruding batten at the top (*b*, fig. 103). The two sides of the hive have two battens, an upper (*U*, fig. 103) and a lower one, marked *L*.

A straw mat is attached to the front of the hive, covering the entire

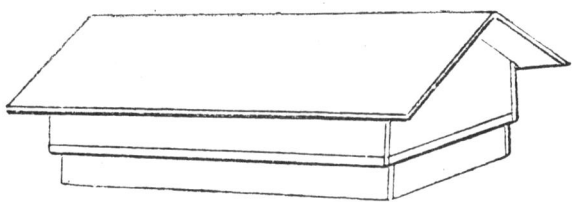

Figure 102. A peaked roof without hinges that can be used in place of a flat one.

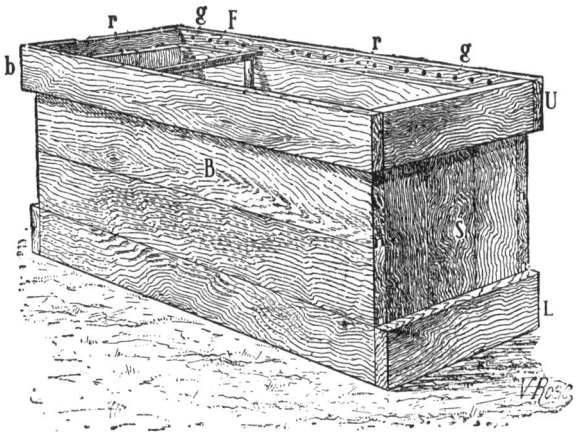

Figure 103. Hive body for a moveable-frame hive. *B*, rear; *S*, one side; *b*, *U*, *L*, battens; *F*, one of twenty frames, in position; *g*, guide marks; *r*, rabbet.

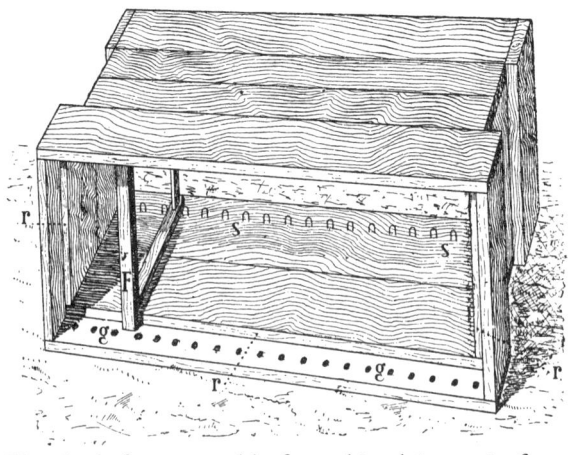

Figure 105. Hive body for a moveable-frame hive, lying on its front: *r*, rabbet; *g*, guide marks; *F*, a frame whose top bar is centered in between two guide marks and whose base is positioned between two staples, *s*. These staples, *s*, line up with the guide marks, *g*.

front, except for the batten at the top and an area 4" (10 cm) high at the bottom (fig. 101). There's another straw mat on the back of the hive that reaches almost all the way to the bottom. This straw helps protect the hive against fluctuations in temperature; it's not needed on the sides of the hive, which are protected on the inside by the combs. A hive furnished with straw in this fashion is less expensive than a hive with a double wooden wall, and provides just as much protection.

The hive's *entrance* (*e*, fig. 101) is located at the bottom of the hive's front, to the left of the area without straw, and it can be narrowed or widened, using a sliding metal tab called the *entrance reducer* (*G*, fig. 101). There's another, similar entrance to the right that can be used instead of the first.*

Now, if we look inside the hive while laying it on its front (fig. 105), we can see two rows of small, regularly spaced protruding staples, *S*, along the bottom of the hive. One of these rows of staples is nailed to the interior of the bottom of the hive's front, and the other is nailed to the interior of the opposite side.

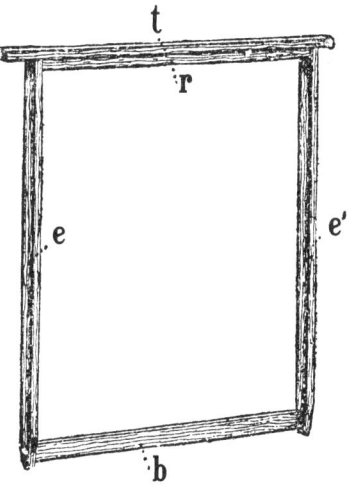

Figure 104. A hive frame: *t*, top bar; *r*, additional crossbar for reinforcement; *e*, end bars; *b*, bottom bar.

Inside, the entire circumference of the top of the hive's body is ringed by a rabbet forming a shelf (*r*, figs. 103 and 105); above this shelf, on the body's interior, are two rows of guide marks (*g*, figs. 103 and 105) that line up precisely opposite each of the staples near the bottom.

Thanks to these staples and guide marks, the position of the 20 frames is clearly indicated. Each frame should be placed such that its bottom is positioned evenly between two staples, and the frame's top bar

---

* In fact, there should be only one functioning entrance at a time, except during winter. The second entrance is only of use when the beekeeper wishes to move a cluster of bees from one side of the hive to the other. In that case, the second entrance should be opened, and the first one closed.

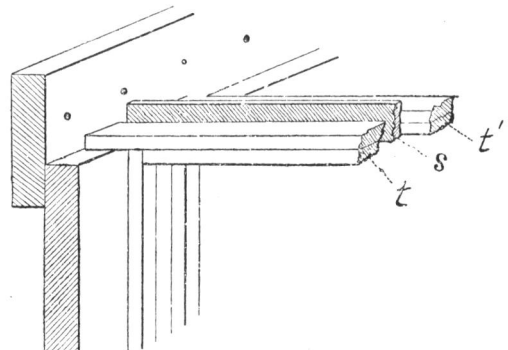

Figure 106. The position of the slats placed in between the frames: *t, t'*, top bars of two frames in position; *s*, slat depicted in gray placed with its narrow side down and protruding above them.

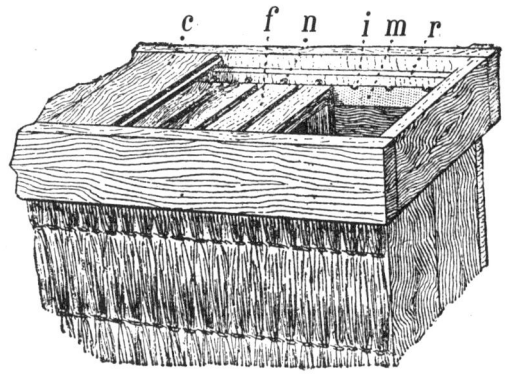

Figure 107. Modifications of a moveable-frame hive to prevent propolis buildup: *f*, top bars of the frames; *m*, strip of sheet metal with notches, *i*; *n*, nail of a top bar positioned in a notch of the sheet metal; *r*, rabbet on which the cover boards, *c*, rest.

(*F*, fig. 105) between two guide marks. Figure 103 shows a frame placed as described, in its natural position.

Once the frames are in place, a gap remains in between their top bars that is covered with wood slats, placed standing on their narrowest side (see fig. 106). Finally, an old wool blanket or a straw mat is placed on top of everything.

The hive's roof consists of four boards assembled into a frame and covered with a thin sheet of galvanized metal, shaded gray in figure 101.

The roof's height makes it easy to place feeders (§ 220) and sections for comb honey (§ 194).

We'll assume that a beginner will acquire, or build himself,* a certain number of hives similar to those we've just described.**

Now, we need to set up these hives in order to install any swarms that may appear.

---

* See G. de Layens, *Construction économique des ruches à cadres* (ed. Paul Dupont).

99. Installing foundation

We've already seen (§ 48) the advantages of using foundation. Beginners, who typically don't have any old comb at their disposal, are advised to install foundation (or at least a bead of wax, as in § 102) in the twenty frames contained in each hive. By the way, since the cost of installing foundation is negligible, you may choose to buy hives with foundation sheets pre-installed by the manufacturer.***

If you'd like to install foundation yourself, here's how to go about it:**** order some foundation sheets whose dimensions are a bit smaller than those of the inside of a frame (in our hive, that would mean sheets 11 7/8" x 14 1/4", or 30 x 36 cm). The sheets are bought a bit smaller than the frame to prevent them from buckling when they expand.*****

To attach such a sheet to a frame, begin by nailing staples to the interior of the frame and in the middle of the top bars, without nailing them in all the way, at the points marked *A*, *B*, *C*, *D*, and *E* (fig. 108); such staples are widely available for sale. To ensure that the staples remain firmly in place, it's a good idea to place them in water with a bit of salt to cause them to rust over.

** You can slightly modify this hive design so the top of the frames is less likely to be tightly glued with propolis. With this modification, you don't place wood slats between the frames, and the bees may build comb in between the top bars, which poses a minor inconvenience.

In a hive set up to prevent frames' top bars from being glued to the hive body (fig. 107) a strip of sheet metal, *m*, is nailed beneath the rabbet, *r*. This metal strip has notches, *n*, to hold rather long nails, *n*, that replace the protruding lugs of the frame's top bar (see figure 109, which shows a frame with its two nails). There are no wood slats between top bars. The rabbet, *r*, is located to the outside of the two metal strips, at a slightly higher level—higher than in the previous hive. In place of the frames, cover boards, *c* (fig. 107), around 4" (10 cm) in width rest on this rabbet, covering the hive completely a bit above the frames.

*** You can often find impure or even artificial foundation for sale at a lower price. It's better to pay a bit more to get pure wax (see § 156).

**** You'll need the following items for this procedure: foundation sheets of the needed dimensions, some staples, a spoon, a bit of wax for melting, a source of heat, and a spur wire embedder or a coin with a groove cut into it.

***** You shouldn't buy foundation that is too thin. Ask for thick sheets; sheets that are too thin can come loose or become warped by the heat inside the hive.

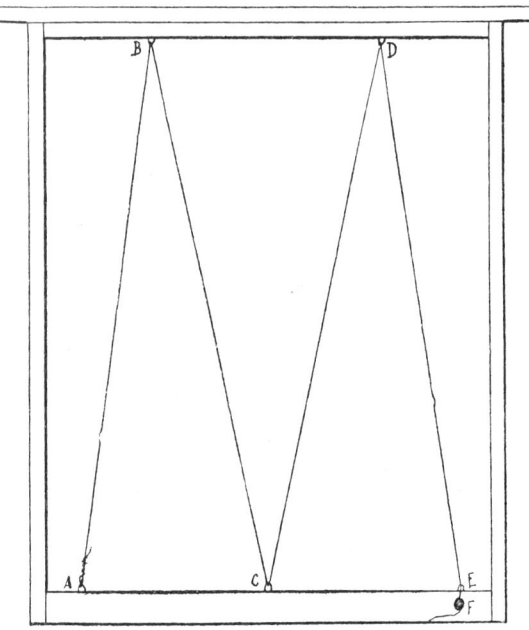

Figure 108. Frame with wire for holding foundation sheet in place.

Take some tin-plated wire, about 24 gauge (0.5 mm thick), and attach it at point A (fig. 108), then move on to points B, C, D, and E. After stretching it in this fashion, wrap the end around an upholsterer's tack, F, nailed into the frame with a hammer. To stretch the wire, use a set of pliers placed as indicated in figure 109, with the upper jaw of the pliers resting on the staple. When the pliers are squeezed, the staple is pressed into the wood, stretching the wire.

Once the wire has been strung in this fashion, lay the frame flat on top of a foundation sheet, which in turn has been laid atop a board that is the same size and the interior of the frame. The wire-strung

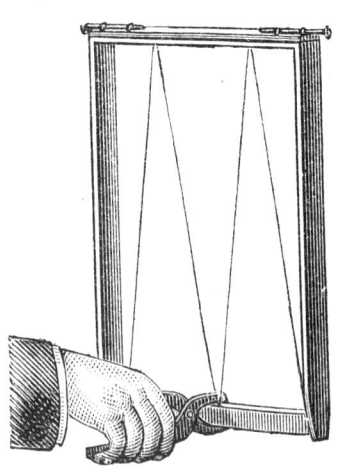

Figure 109. Method for stretching wire onto a frame.

frame should be placed on the wax sheet such that one side of the sheet lines up exactly with the top of the frame, while the other three sides of the sheet should leave a slight gap between it and the wooden bars.

Figure 110. A spur embedder of the kind invented by Woiblet.

Now you need to attach the foundation sheet to the wire. To do so, you can use a small wheel-like device (fig. 110) called a *spur embedder*,* which should be warmed and rolled along the wire, pressing down slightly (fig. 111). Even more simply, you can use a thick coin, into whose edge you have filed a groove; after warming it, slide it atop the wire while holding it with some pliers.

Now that the wire has been pressed into the foundation sheet, tilt the frame so that it rests on the top bar, and, taking a warmed spoon full of molten wax, pour the wax where the foundation sheet meets the top bar. In this way, the foundation sheet will be completely welded to the frame at the top.**

Figure 111. A beekeeper attaching wire to a foundation sheet using the spur embedder.

100. Frames primed with comb

If you've managed to get your hands on some pieces of wax from fixed-comb hives, treated with sulfur and free of brood and drone cells, you can use it instead of foundation. When glued to the top of a frame, these bits of comb can help guide the bees as they build their own comb (fig. 112). If you have some hives that are dead or disorganized after an unsuccessful wintering (§ 86), you can use their comb for this purpose, as long as it has been treated with sulfur first.

101. How to prime frames with comb

Lay the comb flat on a tabletop. Using a thin knife, cut out all of the sections containing drone cells, and trim the remaining pieces until they are straight on one side so that they can be glued to the upper part of an empty frame (fig. 112).

Figure 112. A frame primed at the top with pieces of comb.

For this purpose, use some good hide glue made as follows: place some chunks of glue in a bowl containing four or five times more water than glue, with the bowl in a hot water bath.

When the glue has melted, it should have an oily consistency. Using a brush, cover the side of the comb that you trimmed straight with glue and stick it to the inside of the top bar of a frame turned upside-down (fig. 113). In this way, cover the entire length of the top bar with pieces of comb.

Be careful not to glue the bits of comb together, since the bees will need to demolish any sections that have been glued together in order to build new cells.

* This kind of spur embedder was invented by Woiblet.

** Besides the spur embedder, still in use, foundation is now commonly embedded using 12-volt direct current from a battery or transformer. The current quickly heats the wire, embedding it in the foundation. *Ed.*

Quite often, one can find very affordable comb from hives in which swarms were installed the previous year, only to die out the following winter. In this case, it's a good idea to glue large pieces of comb to the top of the frame instead of simple strips like those seen in figure 112; this will allow the swarm to get down to harvesting honey straight away.

Figure 113. A beekeeper priming frames with pieces of comb.

102. Priming frames with a bead of wax

If you don't have any old comb and can't buy foundation, there's yet another way to guide the bees' comb-building inside a frame: prime the top of the frame with a bead of wax. In order to attach it directly in the middle of the top bar, make a rule in the form of a bracket like that marked *A* (fig. 114); its length should be equal to the internal length of the frame. Set it against the inside of the top bar of a frame turned upside down, marked *B*.

In this position, the rule extends to the middle of the top bar. Make sure to spread tallow on those parts of the rule that will be in contact with the wax to prevent it from sticking. Then, pour some molten wax in the angle formed by the rule and the top bar using a small pitcher

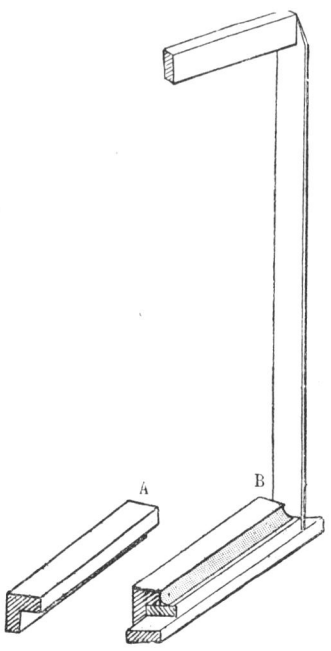

(fig. 116) or an oil can (see also § 221).

Once the wax has cooled, remove the rule, and the frame, restored to its usual position, will now be outfitted on top with a waxen ridge that will help to guide the bees as they build their comb.

Figures 114 and 115. Method for priming a frame with a bead of wax. *A*, a right-angle rule used for placing the bead of wax; *B*, a frame (cross-section) showing the positioning required for pouring a bead of wax to serve as a comb guide.

Figure 116. A beekeeper pouring some wax on the top bar of a frame turned upside down in order to prime it with a bead of wax.

103. Final preparations for installing a swarm

Fill the hive with 20 frames furnished with foundation or primed with comb or a bead of wax. Starting at one side of the hive, insert the frames side by side, making sure that each frame is positioned exactly in between the guide marks and that the gaps between frames are filled at the top with wooden slats, placed standing on their narrowest side.

If you've used foundation, you can use less of it by interspersing foundation frames with simple primed frames.

Now the hive is ready to welcome a swarm.

Since the hive that releases the swarm will need to be moved, you should prepare a stand in advance wherever you plan to locate that hive. Cover the stand with a frame-hive bottom board.

104. How to collect a swarm

We've already seen (§§ 39 and 40) how a natural swarm* is produced and how it leaves the hive.

Recall that swarms will rarely leave their hive unless the temperature is at least 68°F (20°C), and the time is between ten in the morning and three in the afternoon.

When a swarm has been spotted, you need to collect it.** With a veil and a smoker, take an ordinary skep that is totally empty and clean on the inside.

If the swarm seems intent on leaving the apiary, you can toss some sand or ash at it, or spray it with some water from a garden sprayer. You can also fire a gun in its direction or use a mirror to reflect sunlight toward it. Making noise with kitchen utensils, as is often done in rural areas, is of no use whatsoever.

First, let's assume that the swarm has gathered beneath a tree branch. With one hand, hold the skep, turned upside-down, just beneath the swarm when the bees are tightly clustered; with the other, grab the

---

* The term "natural swarm" is used in this book to differentiate it from an artificial swarm (see § 163) or a split (§ 234). *Ed.*

** For this procedure, you'll need: a hat with veil, a smoker, an empty skep, a sheet, and, perhaps, a small broom, a pole and a ladder.

branch and shake it suddenly (fig. 117). The entire swarm will be shaken free of the branch and tumble into the skep.*

Having spread a sheet on the ground in advance, now turn the skep over gently atop the sheet, returning the hive to its usual position, but being sure to hoist it a bit on one side using a small bit of wood. The collected swarm will then fall onto the sheet while initially remaining inside the hive; you'll see a few bees fly away, while the others will exit

Figure 117. A beekeeper collecting a swarm hanging from a tree branch.

the hive in large numbers at the bottom, as if preparing to leave—but they stop abruptly and head back toward the hive.

At this point, you'll see some bees that are said to "make the come-hither signal." Indeed, this signal, made by a general beating of wings, will cause all of the workers to reunite in order to return to the hive.

---

* Today, a 5-gallon bucket or a similar container can be used, covered with a sheet to hold the bees. If using a bucket, do not water it nor turn it upside down until ready to install bees in the hive (see § 107). *Ed.*

The workers making the come-hither signal (fig. 118) stick their abdomen up in the air, instead of lowering it as ventilators do (compare figure 118 to 2, fig. 2). Now, smoke any bees that may still remain on the tree branch to encourage them to rejoin the others. Soon thereafter, the majority of the bees will be reunited in the hive. To prevent the swarm from leaving again, you should cover its hive with several sheets and water it from time to time. Leave it like this until sunset, before installing it in a moveable-frame hive.

Figure 118. Bees producing a "come-hither" signal.

105. When a swarm is awkwardly located

1. *The swarm is hanging from a high branch.* In this case, you'll need two people. One climbs the tree, either directly or using a ladder, while the other holds the skep beneath the swarm—the skep is attached upside-down to the end of a pitchfork or long pole. The person on the tree shakes the branch to cause the swarm to fall into the skep; then, proceed as described above.

2. *The hive has gathered where two thick branches fork, or is spread out along a branch, a tree trunk, or a wall.* Since in these cases it's impossible to shake the swarm into the hive, you'll need another strategy. Use a string to attach the skep above the group of bees, then use some light smoking to guide the bees gradually toward the skep's interior (fig. 119); when almost all of them are there, proceed as described above. You could also use a small broom to cause the bees to fall into a skep placed upside-down beneath the swarm.*

---

* Another option is to place a frame of dark old brood comb near the swarm. The bees will usually move onto the frame and be ready for hiving. *Ed.*

3. *The swarm is simply on the ground, or on a bush.* In this case, cover the swarm with the empty hive, encouraging the bees to enter it with a bit of smoke; then, proceed as described above.

Figure 119. A beekeeper collecting an awkwardly positioned swarm.

106. Determining the swarm's hive of origin

As we know, the swarm will be placed in a moveable-frame hive, and this hive will be located in the same spot as the hive that produced the swarm.

For this to be possible, it's absolutely necessary to know the swarm's hive of origin. If you haven't actually seen the swarm emerge, you can attempt to determine the hive that cast it as follows.

The day after the swarm emerges, you can check to see which hive's activity level has most drastically decreased since the day before—as one might expect, this is the hive that produced the swarm.

Immediately after a swarm has departed, you'll sometimes notice some young bees, pale in color, that have fallen to the ground, unable to follow the others. The hive in front of which such young bees are found is the hive that cast the swarm.

107. Installing the swarm in a moveable-frame hive*

Not long before sunset, place a sheet on the ground, upon which the moveable-frame hive prepared as described above (§ 103) should be set and propped up on one end using a wedge. Then, bring the skep containing the swarm and, with a sharp blow, knock the swarm down onto the sheet in front of the frame hive (fig. 120), and remove the skep that contained the swarm. The bees will head for their new home, enter it, and climb to the top of the frames; you can nudge them a bit by smoking lightly. While the bees are making their way

Figure 120. A beekeeper introducing a swarm into a moveable-frame hive from below.

up the frames, move the fixed-comb hive that cast the swarm to the new stand you've prepared in advance.

After a while, when no more bees remain on the sheet, gently move the moveable-frame hive, which now contains the swarm, onto the stand just vacated by the hive of origin.

---

* For this procedure, you'll need: a sheet, a wooden wedge, a string with a rock tied to one end or a plumb line, a smoker, and a goose feather.

HIVING SWARMS

Be sure that the new hive is *perfectly straight*. All you need to test whether the hive's sides are vertical is a rock tied to the end of a string, used as a plumb line. This is a very important requirement: if the hive is tilted, then the combs—and especially the combs on primed foundationless frames—will be built askew with respect to the frames, even protruding from one frame into another as shown in figure 121. Later, when you go to inspect the hives, it will be difficult to pull out such frames.

The next day, inspect the hive and see what side of it the swarm has gathered on; then, open the entrance on that side, and close the other. Remove the frames that are unoccupied by the group of bees, and place slats atop the area of the hive now left empty.

With these arrangements made, the bees that leave their previous, now relocated hive the next morning, returning naturally to their old accustomed spot, will reinforce the population of the swarm now installed in the new moveable-frame hive.

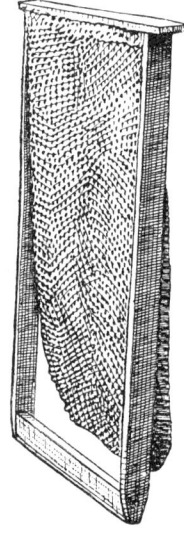

Figure 121. A poorly constructed comb from a hive that was not perfectly level.

This procedure has resulted in:
1. A moveable-frame hive with a large number of bees who, if you've installed foundation or primed the frames with large pieces of comb, already have comb at their disposal that is almost ready for the foraging season.
2. A relocated parent hive whose population is now smaller, but which still has a good supply of honey, a virgin queen, and plenty of brood to replace the bees it has lost.

The procedure we've now completed has yet another advantage: the hive of origin, now relocated, will be much less likely to produce an afterswarm (§ 41), which is always best avoided (§ 111).

However, be sure to take note of the date of these procedures just in case of an afterswarm.*

CHAPTER 7. MOVABLE-FRAME HIVES. SWARMS

Installing a swarm through the top of a moveable-frame hive. You can also install a swarm through the top of the hive. In this case, you should proceed as follows.

Place only ten to twelve frames, then, with a sudden blow, shake the bees from the basket into the empty area of the hive (figure 122). Then, cover the hive with a sheet to prevent the bees from flying away, and, using a smoker, send some smoke up underneath the sheet while standing on the same side as the empty part of the hive where you deposited the bees—this will encourage them to move onto the frames. Then, open the hive entrance on the same side as the frames, and leave the other one closed. The next day, make sure the frames haven't been disturbed during the procedure.

Figure 122. A beekeeper introducing a swarm into a moveable-frame hive from above.

* If, when carrying out this procedure, you already have some populated moveable-frame hives, it would be a good idea to add a brood frame from a strong hive to the swarm, which helps prevent the swarm from absconding—a rare event, by the way.

108. What if you're unable to identify the swarm's hive of origin?

It may happen that you are unable to tell which hive produced a given swarm. In this case, you'll have to give up the idea of placing the new frame hive containing the swarm in the location of its parent hive, and be content with simply placing it on the stand prepared in advance to receive the hive of origin.

109. Feeding a swarm in the event of bad weather

As we know (see the end of § 40), swarming bees are full of honey, and their wax glands, ready to get to work, produce numerous scales of wax for the bees to later fashion into comb.

So as soon as the swarm has been placed in the moveable-frame hive, the bees will begin to draw out the foundation sheets or the bits of old comb used to prime the frames.

However, if the weather is bad, the honey the bees are full of may not be enough to last several days. In this case, you'll need to help out the swarm by giving them some sugar syrup to tide them over until the weather improves.

If you don't have a feeder (§ 220), you can simply put a bowl full of syrup (§ 89) on the bottom of the frame hive, in the empty area—be sure to add some slices of cork or some bits of straw on top. Push the bowl until it touches the first frame, then, using a goose feather, sweep a few bees from the swarm down onto the syrup. Be sure to remove the bowl the next morning.

We should add that even in good weather feeding a swarm may help it get off to a good start.

110. What to do when there is an afterswarm

You've taken note of the date on which the primary swarm emerged. If the good weather lasts, then an afterswarm will generally emerge eight or nine days later, if at all. A beekeeper can be warned of a coming swarm a day or two in advance by the queen's song (§ 41). So beginning on the fifth day, it's a good idea to listen in the evenings to see whether the hive's queens are singing.

If by the tenth day there is no singing in the hive, then it will not produce an afterswarm.

Let's assume that you've heard the queen's song. This means that an afterswarm is on its way, and is sure to emerge within the next day or two, as long as the weather is good.

111. Collecting an afterswarm

When the emergence of the afterswarm has been signaled in advance, you should prepare to collect it—not to install it in a new hive, but to put it back into its hive of origin. This swarm is much smaller than the primary one, and will most likely not have enough time to gather a honey stockpile for the winter; moreover, their hive of origin no longer has a large enough population itself. And as we'll see, one of beekeeping's cardinal rules is to *always maintain strong hives*. So the afterswarm will need to be put back into the hive that produced it.

Generally speaking, it's more difficult to collect an afterswarm than a primary swarm, because its queen—a young one—is better at flying, allowing the swarm to go higher and further. You should collect it in the same way as the first (§ 104).

Once collected in the empty skep, the swarm should be covered in a wrapping canvas and moved into a cellar, having propped it up on one end with a wedge. It should only be returned to its parent hive the next evening, in order to prevent it, as much as possible, from leaving the hive once again.

112. Determining the afterswarm's hive of origin

Just as you did with the primary swarm, you should determine the afterswarm's hive of origin in order to know which hive it should be returned to. In fact, it may happen that several primary swarms emerge one after the other, and that several hives have the potential to produce afterswarms—and this is when it becomes important to know which relocated hive produced the afterswarm. A hive that has swarmed and that hasn't been relocated will be more likely to produce an afterswarm.

1. If you've already heard the queen's song from all the hives in question, then all you need to do is listen for it on the evening of the day the swarm emerged—the hive where you no longer hear the song is the hive that produced the afterswarm. And this is the hive to which the swarm should be returned.

2. If you haven't listened for a queen's song, there's still another way to determine the afterswarm's hive of origin. The morning after, lightly smoke the swarm collected in the empty hive, and use a ladle to scoop up some bees and toss them into a small dish full of flour. Carry the bees, now white with flour, several steps away and let them fly away. Now, watch the entrances of the various hives that you suspect may have produced the swarm—the hives you see the whitened bees enter is the afterswarm's hive of origin.

113. Returning an afterswarm to its hive of origin*

The next evening, move the swarm from the cellar next to the hive that produced it; lightly smoke the hive of origin, remove the wrapping canvas, and, having placed two sticks of wood on a sheet, abruptly dump the bees of the swarm onto the sheet in between the two sticks. Then, gently place the hive of origin onto the sticks, above the bees, and smoke all around the bees to encourage them to gather in a single cluster.

This procedure for returning an afterswarm to its hive of origin has two major advantages:
1. It avoids both preserving a swarm that is still too weak, and reducing the strength of the hive of origin.
2. Reduces the possibility of the afterswarm leaving its hive again. In fact, it's all but certain that the swarm will not leave again once it's been reunited with its former hive.

114. Various circumstances you may encounter when swarms emerge

1. *The swarm returns to the hive that produced it.* Due to a change in the weather, an entire swarm may return to its hive of origin; in this case, you should expect to see it reemerge on the first day of nice weather.

---

* See also § 233.

2. *The swarm enters another hive.* A swarm, once emerged, may sometimes fall upon another hive. If this happens, you'll see a battle break out between the bees of that hive and the swarm. If possible, you should swap the location of the besieged hive with that of the hive of origin—eventually, calm will be restored. If it's too late to switch the hives, blow some smoke in the thick of the fighting until it breaks up.

3. *The swarm's queen has gotten lost.* You may see a swarm whose bees are flying around aimlessly for a very long time, without forming a compact group. This indicates that the swarm has lost its queen. You may find the queen on the ground in front of the hive, often surrounded by a small group of worker bees; in this case, you can pick her up and put her under a glass. Remove the hive that produced the swarm and place it on the empty stand you prepared in advance. Then, set the glass with the queen underneath on the bottom board that is now empty. Atop the queen, place the empty skep meant to collect the swarm, after gently removing the glass.

All of the bees that left the hive will now soon return to it. In the evening, you can transfer the swarm into the moveable-frame hive as indicated above (§ 107).

4. *Two or three swarms emerge at once.* Several swarms may emerge simultaneously. Often, two or three swarms may gather in the same place, forming a single cluster; or, less frequently, a second swarm may join a swarm that has already been collected in a hive.

To keep things simple, don't even try to separate the swarms; this way, you'll end up with a single queen and a very large population formed by the combined swarms, which you can install in a single moveable-frame hive.*

5. *A swarm from a swarm.* When the swarming season is prolonged, a swarm that has been installed in a fixed-comb hive may itself

---

* If you already have some populated moveable-frame hives, it's a good idea to give these united swarms a frame containing brood, which will help prevent them from leaving.

sometimes produce a swarm before the season ends*—but this will almost never happen with a moveable-frame hive.

115. The apiary after swarming season

What does the apiary look like once swarming season is over?

Let's assume that we began with five hives (*A, B, C, D, E*, fig. 123), and that three of these hives yielded a primary swarm. If we dealt with each hive that produced a swarm in the manner indicated above (§ 107), then our apiary will now consist of:

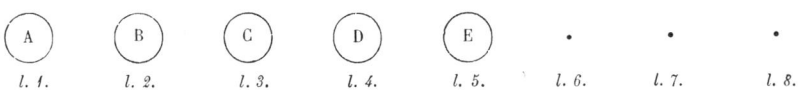

Figure 123. The status in spring of an apiary consisting of 5 fixed-comb hives (*A, B, C, D, E*) in locations *1, 2, 3, 4, 5*. A number of hive-less stands have been positioned in advance at available locations such as *l. 6, l. 7, l. 8*, etc.

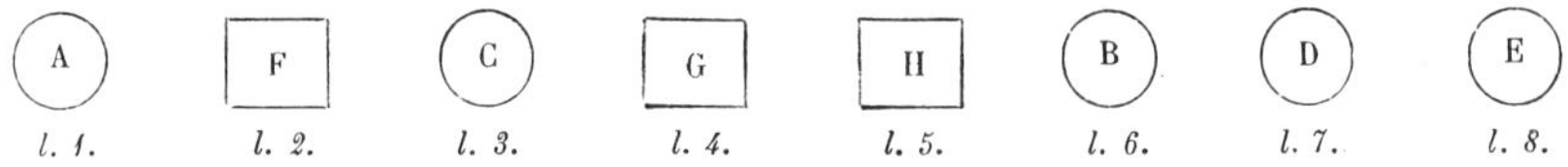

Figure 124. The status of the same apiary after swarming season, if three of the fixed-comb hives have swarmed.

A, fixed-comb hive that has not swarmed, and remains at location 1.
B, a fixed-comb hive that has swarmed, and has been moved to a new location, 6.
F, a moveable-frame hive, now housing the swarm from hive *B*, and placed at location 2 previously occupied by hive *B*.
C, a fixed-comb hive that has not swarmed, and remains at location 3.
D, a fixed-comb hive that has swarmed, and has been moved to a new location, 7.
G, a moveable-frame hive that now houses the swarm from hive *D*, and has been placed at location 4, previously occupied by hive *D*.
E, a fixed-comb hive that swarmed, and has been moved to a new location, 8.
H, a moveable-frame hive that now houses the swarm from hive *E*, and has been placed at location 5, previously occupied by hive *D*.
Thus, in autumn the apiary consists of 3 moveable-frame hives and 5 fixed-comb hives, of which 3 have been relocated.

---

* Such a swarm is called a maiden swarm.

1. Two fixed-comb hives (*A* and *C*, fig. 124), which have not produced a swarm and remain in their original locations.
2. Three fixed-comb hives (*B*, *D*, *E*, fig. 124) that have produced a swarm, but have either not produced afterswarms, or have had their afterswarms returned to them. The hives have been moved, and now stand in their new location (*l. 6*, *l. 7* and *l. 8*, fig. 124).
3. Three moveable-frame hives (*F*, *G*, *H*, fig. 124) containing the three primary swarms; these three hives now occupy the spots previously occupied by the swarms' three hives of origin.

Summary

Preparing moveable-frame hives
Once the bees have begun collecting nectar in large quantities, you should begin preparing moveable-frame hives to receive swarms. In each hive, place twenty frames: foundation frames alternating with frames primed with old comb; or frames with old comb exclusively.

Swarming
When a natural primary swarm is produced, you can collect it in an empty skep, sprayed with water from time to time and left in this condition until evening.

If you've successfully identified which hive produced a swarm, then you can install the swarm in a frame hive readied in this way, a bit before sundown; then, move the hive of origin to a new stand prepared in advance, and place the new moveable-frame hive that has received the swarm on the stand just vacated by the hive of origin.

If the weather is bad after the swarm is installed in the frame hive, you'll need to give it some sugar syrup.

Around eight days after the primary swarm emerges, the same hive may yield an afterswarm, presaged by the queen's song.

When this second swarm emerges, collect it and return it to the hive that produced it.

Where the apiary stands after swarming season
If, for example, a beginning beekeeper has five hives, and three of them produce a primary swarm, then at the end of the swarming season he will have:
1. Two fixed-comb hives that have not produced a swarm.
2. Three fixed-comb hives that have produced a swarm and have been relocated.
3. Three moveable-frame hives containing three primary swarms, now located where the three hives of origin used to be.

Chapter 8
What to Do During The Summer of Year One

116. Handling an empty moveable-frame hive

Before inspecting a moveable-frame hive with bees inside, a beginner is advised to practice handling some moveable frames in an empty hive. In this hive, place ten or so frames, making sure that the bottom of each is positioned precisely in between the protruding staples, and the top bar in between the corresponding guide marks.

With the ten frames placed to one side of the hive, take the one closest to the empty space and practice moving it by placing it one or two slots further over. Also, practice tilting a frame that is in place by leaving the bottom in between the same staples and moving only the upper part of the frame.

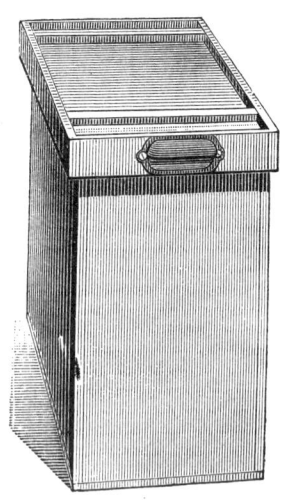

Figure 125. A frame box.

117. A frame box

When you inspect a moveable-frame hive, you'll often need to add or remove frames. To do so, use a wheelbarrow to transport, along with the tools you need, a box that can hold several frames (fig. 125). This *frame box* should be tightly sealed to prevent the bees from getting into the box when it contains frames. A rimmed tin tray should be placed in the bottom of the box to catch any honey that may drip down from the frames.

118. Inspecting moveable-frame hives*

Let's inspect a moveable-frame hive on a nice afternoon approximately ten days after installing a swarm there. Open the hive after lightly smoking it near the entrance with a handheld smoker to drive

Figure 126. A beekeeper smoking a moveable-frame hive.

back the guards (fig. 126), then remove, one by one, the slats between the top bars or the cover boards covering the body of the hive on the side of the hive opposite the open entrance—that is, on the side where there are no frames. As we go about removing the slats or cover boards, blow some smoke from top to bottom into the empty space.

---

* For this inspection, you'll need: a smoker, a knife, some goose feathers or a bee brush (see fig. 128), a veil, and a frame box. It's a good idea to have an assistant for this operation.

Now we've made our way to the first frame—on which, at this point, little or no comb will likely have been built.

Having removed the slat separating this frame from the next one, and leaving the bottom of the frame in position between the same staples, let's tilt the upper part in the direction of the empty space.

For added force, you can use a tool called a *frame grip* (fig. 127). This is a kind of clamp that grabs the top of the frame and can be easily operated by hand. When you're inspecting frame hives that have been populated by bees for an extended period, the frame grip will also allow you to pry loose a frame that has become stuck with propolis (§ 18).

When you tilt the first frame, smoke the bees a bit, again moving from top to bottom, making sure that the smoke penetrates the gap, now widened, between the two first frames, and continue to smoke until the bees are pacified (§ 58), producing a loud, clearly audible buzzing sound.

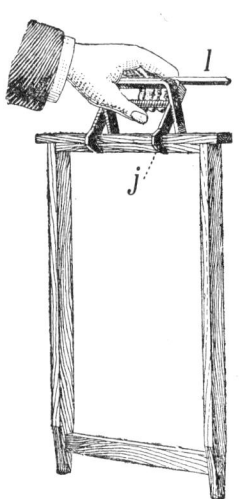

Figure 127. A frame lifted using a frame grip; *j*, one of the jaws; *l*, the lever.

Now, gently lift the second frame (fig. 128) after removing the slat separating it from the third, tilting a bit toward the first to avoid crushing the bees. If the workers have begun working beneath the bits of comb used to prime the frame (or, if foundation sheet was used, to build comb atop the stamps indicating the shape of the cells), then you'll see some liquid honey in the upper cells. Put the frame back in place, then tilt the upper part toward the first frame in order to inspect the third. Once again, blow some smoke into the newly formed gap, and again into the first. Inspect the third frame just as you did the previous one—you don't even need to remove it all the way.

Two consecutive frames may be joined in several places by newly built comb, especially if you primed some of your frames with pieces of comb. This is no real cause for concern—use a knife to carefully cut through this brace comb, then pull out the frame for inspection.

CHAPTER 8. SUMMER OF YEAR ONE 137

During this inspection, take care to make sure that all the combs are being built straight inside each frame—this is extremely important when it comes to handling moveable frames.

If all of the frames had sufficiently thick foundation that was properly installed, then generally what you'll see is comb construction proceeding regularly within the frames.

Figure 128. A beekeeper inspecting a frame from a moveable-frame hive.

If the frames were primed with bits of comb only, or if the foundation installed was too thin, then it may be the case that one or more combs have been built askew inside their frames; in this case, the comb within a given frame may be warped, with bumps and other irregularities.

In this case, remove the irregularly built frame entirely and place it in area of the hive with no frames, then put the cover back atop the section with the frames.

Then, drive the bees off of the irregular frame, using a goose feather or a bee brush (figs. 129 and 130) to sweep them off the comb toward the bottom of the hive. Then, carefully straighten out the warped sections of comb by hand, trimming them with a knife if necessary. Put the frame back into its original position, and continue your inspection.

Figure 129. A bee brush.

Figure 130. A beekeeper using a brush to drive away bees covering a frame.

As we inspect the various frames during this procedure, we should be able to easily recognize young brood—that is, eggs and young larvae (e and $y.l.$, fig. 36), and, almost always, some honey in the upper portion of the comb.

You'll also find some honey in the top of frames with comb containing no brood (fig. 131).

If the first frame you inspect—that is, the one nearest the empty area of the hive—is already well advanced in terms of comb construction and contains some honey, then you'll need to add two or three more frames primed with foundation or, if none is available, with pieces of comb.

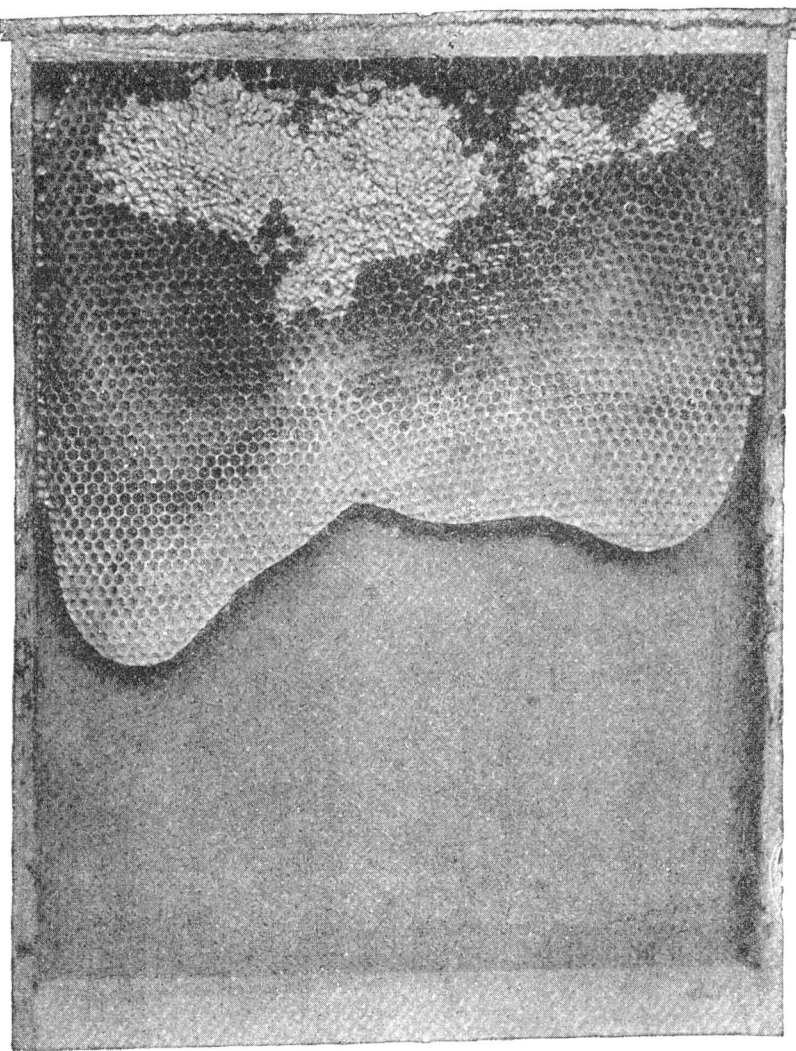

Figure 131. Comb under construction in a frame primed with pieces of old comb attached to the top of the frame. You can see the new, white wax nearer the bottom of the comb, and, at the top, an area of cells containing honey, already partially capped.

You may sometimes find that the bees have begun constructing comb on the boards covering the empty area of the hive. This indicates *with absolute certainty* that the colony doesn't have enough frames and that some must be added.

As you finish this initial inspection, be sure to put all of the frames and slats back in place and close the hive.

119. The advantages of foundation frames when a swarm is installed in a moveable-frame hive

When swarms are installed in a moveable-frame hive, during a heavy honey flow, you may find more honey in hives containing foundation or primed with large pieces of comb than in those with only a bead of wax at the top of the frames.

This is due to the fact that the bees are able to build comb more quickly in the former instance than in the latter.

120. Monitoring the remaining fixed-comb hives

If after swarming season the bees are still gathering honey, it's a good idea to see whether any fixed-comb hives that haven't swarmed are now too small for their colony size.

In this case, after smoking the fixed-comb hive through the entrance, tilt it in order to inspect it from the bottom. You'll know that there's not enough room if you see a large number of bees gathered on the bottom board—and if, when inspecting the combs, you see brood cells near the very bottom.

When you've determined that a colony has outgrown its hive, you can expand it using a kind of straw or wicker cylinder called a *super*, which is placed beneath the hive (fig. 132), attached using wire hooks.

Placing a super beneath a fixed-comb hive has another advantage: in regions with an early-autumn honey flow, the super will usually prevent the hive from producing a late swarm that would have trouble gathering enough provisions for winter.

121. Monitoring moveable-frame hives

If the bees are continuing to forage, or if you are in a region with abundant heather or buckwheat, it's a good idea to complete another inspection of the moveable-frame hives where swarms were installed, to see whether you need to add more frames primed with foundation or pieces of comb.

Figure 132. A beekeeper placing a super beneath a fixed-comb hive.

122. End of the honey season

The less nectar the bees are collecting from the flowers of late summer and autumn, the fewer eggs the queen will lay. Many bees die, and are not replaced by new ones. The hive population will drop, and the remaining bees will begin clustering, occupying fewer of the combs.

The following signs indicate that the season is over:
1. The bees' activity level around the entrance has dropped off, even when the weather is nice.
2. The drones have been expelled by the workers—either chased away or killed.

3. If you weigh the hives, you'll see that their weight is not increasing, or has even begun to decrease.
4. You no longer see any fanning bees near the hive entrance.
5. Here and there, you notice bees roaming around the hives, trying to get in; these are robber bees, ever greater in number during this time of year.

Summary

Inspecting moveable-frame hives

After learning to handle an empty moveable-frame hive, a beginner should inspect the hives approximately ten days after installing swarms.

During this inspection you should:
1. Check to see whether the combs are being built straight within their frames; if not, straighten them out or remove any irregularly-built areas.
2. Make sure that each hive has brood of all ages.
3. See whether comb construction is at an advanced stage inside the frames, and whether you need to add more frames, primed with foundation or pieces of old comb.

Monitoring the apiary

Next, inspect the fixed-comb hives still remaining in the apiary, and if any of them that haven't swarmed are becoming too small for their colonies, you'll need to add a super beneath the hive.

Later, continue to monitor the moveable-frame hives. If an autumn honey flow is underway, the beginner would do well to inspect the hives once again to see whether more frames should be added.

Chapter 9
What to Do During The Fall of Year One

123. How a beekeeper harvests honey

A beekeeper should harvest surplus honey from the hives before seeing the signs indicating the end of the season. If you put it off for too long, the bees will become more difficult to handle, since, as we know, they grow more irritable when the flowers have run out of honey. If the apiary is in the condition we've assumed, then a beginning beekeeper shouldn't expect a significant harvest at the end of the first year.

Since you'll need to transfer any remaining fixed-comb hives into frame hives the following spring, it's simpler not to harvest any honey from them.

As for moveable-frame hives containing this year's swarms, a harvest can only be expected following an unusually bountiful honey season.

Still, unless the season was exceptionally poor, a beginner should be able to harvest a few combs of honey, at least two, in order to gain a practical understanding of how honey is harvested from moveable-frame hives.

124. Inspecting hives at the end of the season and evaluating the weight of honey in a frame

While harvesting a few frames of honey in order to learn how to extract it, a beginner should conduct a full inspection of the moveable-frame hives after the foraging season is over to determine their condition, and to find out how much honey they contain, in order to be sure to leave enough honey for the winter in each.*

In order to avoid possible robbing, you should conduct this inspection late in the day, shortly before sunset, and make sure to narrow the entrances of all the hives, including the fixed-comb hives, until only two or three bees could pass through at a time.

Let's start by inspecting the strongest and most active moveable-frame hive. We'll proceed as described in § 118, but we'll need to use a lot of smoke, especially if it's very late in the season, since the bees tend to be more irritable then.**

Tilt the frames and begin with the one closest to the empty section of the hive. In general, you'll find varying amounts of honey in the frames; if we find one whose comb is completely capped, it will be easy to estimate its weight without using a scale. Since 20 square inches of capped honey (including both sides) contain roughly 1 lb of honey (1 kg per 300 cm^2), it follows that a full frame in the kind of hive we're using will contain approximately 8 lb (4 kg) of honey.***

125. How much honey should be left for winter reserves?

The amount of honey that should be left to a hive for wintering is a very important question in beekeeping. If you don't leave your hives sufficient honey reserves, they may not survive the winter for lack of honey, and even if the colonies are still alive when winter is over, you'll often have to feed them during the spring.

So resist the temptation to harvest too much honey; *always leave the bees more than enough honey for the winter.*

---

* For this inspection, you'll need: a frame box, a knife, a smoker, a veil, and a feather or bee brush. It's good to have an assistant for this procedure.

** Sometimes, the bees of a certain colony may prove unusually irritable. If this is the case, don't be afraid to smoke them heavily and for an extended period; you can also pour some sugar water in between the frames using an oil can—this will calm them down considerably.

*** This amount will be lower if the comb is very old, since in this case the empty comb will weigh more (§ 30).

Since, if spring comes late, the bees won't be able to gather nectar the following year until the end of May, it's prudent to leave at least 35 lb (16 kg) of honey in each hive.*

It follows that you should first evaluate the total amount of honey a hive contains before removing some frames from the hives we inspect.

You'll quickly get the hang of estimating the approximate weight of each frame, assuming a weight of 8 lb (4 kg) when the comb is thicker than the frame, and is completely filled. We'll judge by sight the surface taken up by honey in each comb, and, in turn, its approximate weight.

We'll take advantage of this inspection of all the frames to confirm that some of them still contain brood, which tells us that a colony still has its queen and is doing well.

In § 131, you'll learn what to do if you find no brood, which, if it's not yet late in the season, signals that the colony has become queenless.

Once we've determined the total weight of the honey, we'll know how much we can harvest from the honey frames. Of course, we'll be sure only to harvest from frames containing only honey, with no brood. Proceed as follows.

Continuing to smoke heavily, completely remove the honey frames that you want to harvest; move them, along with the bees covering them, to the empty section of the hive. Then, move the remaining unharvested frames over until there is no gap in the row of frames, then put back the slats in between each of the frames. Using a goose feather, while continuing to smoke, sweep the bees to the bottom of the empty area. In this way, remove each honey frame, put them in a frame box, and close the hive.

Treat each of the other frame hives in the same fashion.

If we find that a hive's total estimated honey is less than 35 lb (16 kg), then we'll add honey instead of harvesting any. This is done very simply, thanks to our movable frames, since all we need to do

---

* This amount would be needed for Layens hives in northern France (Zone 6) and may require adjustment based on your local conditions: severity and length of winter, race of bees, hive model, management style, and colony strength. Check with local beekeepers to see what is suggested for your area. *Ed.*

is add one or several frames taken from a strong hive to one whose reserves are low, to fill out its winter provisions.

126. When the moveable-frame hives are short on honey

After a bad year, our moveable-frame hives may not have enough honey to safely survive the winter, much less for harvesting.

In this case, it's wise to give any hives that are running short the several pounds of reserves they need, in the form of sugar syrup, to reach a total of 35 lb (16 kg) of honey.

It's very important to do this as soon as possible, since if you wait too late in the season, a drop in the outside temperature may prevent the bees from capping the syrup, which could lead to a poor wintering.

127. Feeding moveable-frame hives*

We can proceed as described previously (§ 109), or, better yet, in the following way, which is preferable during this time of year.

Make syrup by dissolving, under heat, some sugar in water, in a ratio of 14 lb of sugar per 1 gal of water (5 kg per 3 L). Once the water has begun to boil, allow the syrup to cool and pour it into an oil can.

When colonies are fed, a certain amount of syrup is always used up by the bees amidst the excitement caused by feeding. Indeed, the bees' behavior inside the hive is the same as during foraging season—leading to a warmer interior temperature, the raising of new brood, etc.

So, generally speaking, calculations have shown that one should increase the amount of sugar one wants to give the bees by a quarter.

If, upon inspecting a moveable-frame hive, you find that it should be fed in order to fill out its stockpiles, then carefully remove some built-out frames that contain no honey, or very little.

These empty or near-empty frames should be moved to an enclosed room, beyond the bees' reach, and filled with syrup in the following way: place a frame flat on an oilcloth spread on a table, and, using the oil can, pour some syrup in these empty cells, then put a sheet of paper on the frame, whose first side has now been filled with syrup,

---

* See also §§ 220 and 232.

and turn it over on top of the same oilcloth. Fill the second side of the frame in the same way, and, after removing the paper, put the frame—now full of syrup—back into the moveable-frame hive. This is easily done without spilling the syrup inside the cells, since the ratio of sugar to water suggested above will produce a syrup thick enough to resist spillage under these conditions.

The syrup-filled combs should be distributed, in the required number, to each hive. Only place them in the evening, around nightfall, to avoid robbing.

In strong hives, you can add as much as 11–13 lb (5–6 kg) of syrup at a time.

Figure 133. A beekeeper pouring some syrup in the cells of an empty frame.

128. What if robbing breaks out?

We've already detailed what should be done with fixed-comb hives when robbing breaks out due to an oversight by a novice beekeeper.

You can do the same with moveable-frame hives if the robbing is particularly intense.

But, when the robbing is just beginning, it's simplest to handle moveable-frame hives in the following manner.

Place the smoker in front of the entrance of the hive that has begun to be robbed (fig. 134), which will prevent bees from entering. The robbers will leave the hive one by one, and be unable to return; a half-hour later, remove the smoker and narrow the entrance such that only a single bee can enter at a time. Then, you can sprinkle the outside of the hive with a bit of kerosene—except for the entrance. These precautions will usually be enough to nip robbing in the bud.

Figure 134. Position of a smoker in front of a moveable-frame hive when robbing has broken out.

129. Tools required for harvesting honey from moveable-frame hives

We've seen (§ 47) how an extractor works—and the extractor is the main instrument for harvesting honey from frames without destroying the combs.*

---

* Don't forget, in advance, to use some bicycle oil on the various extractor components subject to friction. (Food grade grease is recommended today. *Ed.*)

But one difficulty presents itself: as we know, honey in its finished state—that is, the state in which it can be preserved without fermenting—is stored in cells that have been capped by the bees. So, ideally, we need to harvest combs whose cells have all been capped—or, failing that, combs whose cells are at least two-thirds capped. This means that we'll need to remove the caps before putting the frames into the extractor.

For this purpose, you can use an *uncapping rack* for holding a frame that needs to be uncapped, and a special knife for removing the caps.

An *uncapping rack*, made of wood, is designed to hold the frame at the most convenient angle (see fig. 137).

There are two hooks at the top of the rack for holding the two ends of the top bar of the frame.

Held down in this way at the top, the frame filled with honey rests on the rack in the position required for uncapping.

The best *uncapping knife* is a two-handled knife (fig. 136) whose blade is slightly cupped, with the cutting edge on the bottom; this configuration allows the majority of detached caps to be removed without becoming stuck again to the parts that have already been cut.*

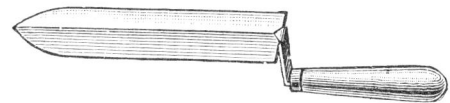

Figure 135. An uncapping knife with a single handle.

Figure 136. An uncapping knife with two handles.

A container (for example, a tin tub) is placed at the foot of the rack, in between and a bit in front of the two posts, covered with a sieve to catch most of the caps, along with the honey attached to them.

---

* You can also use a single-handle knife (fig. 135), but things will take longer.

A heat source of some kind should be kept near the tray, for warming the blade of the uncapping knife in order to ease the procedure (fig. 137).

The honey that comes out of the valve at the bottom of the extractor will contain a certain amount of wax debris—it will need to be *filtered*.

A *honey settling tank* is, quite simply, a container, much taller than it is wide, with a small hole at the bottom that can be plugged with a cork, or a spigot.

You should also have some containers to hold the harvested honey. The best and lightest containers are hermetically sealable tin cans like that shown in figure 138. They are widely available for sale, in many sizes.*

Figure 137. A beekeeper uncapping a frame hung from two hooks. When he reaches the bottom of the frame, he'll clean the knife and replace it with another left for heating on a small stove.

---

* Today glass jars are preferred. Glass is non-reactive and will not change the flavor of honey. *Ed.*

130. Extracting honey

If, as a beginner, you've harvested at least two frames, then you can practice extracting the honey from them using an extractor. This device should be kept in a room sealed off from bees, where you can bring the honey frames you've taken from the hives.

Take one of these frames and put it on the uncapping rack readied beforehand, and heat the uncapping knife until you can no longer touch it with your fingers. The knife's blade is made slightly shorter than the frame's interior for ease of handling. When the knife has reached the desired temperature, move it from top to bottom to remove the portion of the comb that protrudes beyond the frame's end bars. The caps, along with some honey, will fall onto the sieve placed at the foot of the uncapping rack. Use a spoon to scrape off the knife blade before putting it back on the furnace for reheating.

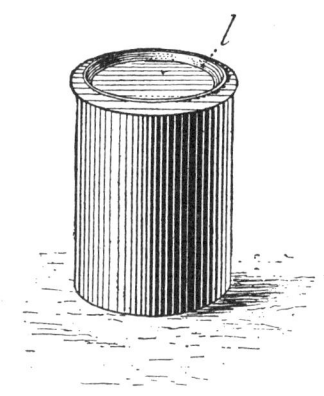

Figure 138. A honey tin; *l*, lid.

If in places the comb is somehow indented or irregular, use an ordinary knife to uncap any cells that eluded the uncapping knife. Then, turn the frame over on the uncapping rack and repeat this process with the other side.

Since comb that was built onto the pieces of comb used to prime the frame instead of foundation may be fragile, it's a good idea to place wire mesh with 2" (5–6-cm) cells atop each side of the frame; the two meshes aren't attached to the frame, but simply joined at the top with two strings. With the frame thus arranged, place it inside the cage in the extractor (fig. 139). Do the same with another honey frame, being sure that the two frames you've chosen and placed on opposite sides of the extractor are of approximately the same weight, to ensure that the device spins evenly during operation.

Now that everything is ready, turn the crank on the extractor, rather gently at first to avoid breaking the comb. Centrifugal force throws the honey out onto the walls of the extractor, where it flows

Figure 139. A beekeeper placing an uncapped frame, protected by a screen, into an extractor.

downward.* You'll hear a sound very much like rainfall; a moment after this sound has stopped, turn the frames around in order to extract the honey from the other side. This time, we can turn the crank a bit more quickly, and for a longer time, to be sure to extract all of the honey from this side of the frames. Finally, we can turn each frame around once again and turn the crank rapidly to extract any remaining honey from the first two sides of the frames.

Now, remove the frames that have been extracted, and place them in the frame box; you can return them to a hive one evening in the next several days to allow the bees to clean them.

Of course, if you have enough frames to harvest, you can put four at a time in the extractor.

---

* If the frames are filled with heather honey, you won't be able to extract it directly using the extractor. In this case, see § 167.

From the spigot at the bottom of the extractor, draw the honey and pour it into the settling tank, along with the honey in the container placed at the foot of the uncapping rack. As for the caps wet with honey remaining on the sieve, you can press them with a spoon to squeeze the honey through the sieve, then toss them into a container—if you have enough, you can put them to use as detailed below in § 264.

Once all the bits of wax rise to the surface of the honey in the settling tank—which may take some time—you can draw the honey and put it in jars for storage.

Since honey easily absorbs moisture, the jars it is stored in should be kept in a dry and well-ventilated place, unless they are hermetically sealed as detailed in figure 138.

131. Hives that are queenless or almost without honey

During the fall inspection, certain moveable-frame hives may be found to have a serious shortage of honey—for example, less than 18 lb (8 kg)—and you have no honey frames to give them. In this event, it's a good idea to unite any such hives into one, especially if a colony seems too weak to store and cap the large amount of syrup you'd need to feed it.

The same is true for any queenless moveable-frame hive: it will need to be united with another.

132. Uniting moveable-frame hives*

To unite two colonies into one, do the following, always late in the day right before evening.

Open the hive you want to unite with another, and lightly smoke it in between the frames; then, remove the frames one by one, covered

---

* The items you'll need for this procedure include: an oil can that can hold around two cups of heavily sweetened water flavored with a drop of essential oil of anise, mint or some other scent; a frame box large enough to contain all of the frames from the hive you're planning to unite with another; a goose feather or a bee brush; a smoker and a veil. (See also another procedure in § 235).

with bees. Pour a bit of scented sugar water on both sides of each frame, then put them in the frame box. The bees will feed on this sugar water, and will take on the same scent we'll soon give to the hive meant to welcome them—this simple trick will ease the two hives' unification and prevent the bees from fighting.

Once all the frames have been moved in this way to the frame box, along with the bees that are on them, close the frame box and move it near the hive they are about to join.

Remove the top of the host hive, blow some smoke in between the frames as you remove the slats and replace them, one by one; then, pour some sugar water into each gap—use around a glassful total for the entire hive.

Move the frames in the hive aside until you reach the first that contains brood. Open the frame box and take out the brood frames from the hive being added, and place them alongside the brood frames of the receiving hive. Then add the frames that contain the most honey, followed by those with less. As for the frames with no honey, sweep any bees that are on them into the new hive and put the frames back into the frame box.

Now, return to the old hive, where a few handfuls of bees still remain.

Lift the body of the hive, causing the bees to fall onto the bottom board; move the bottom board atop the hive where you've just put the frames, and shake the bees down into their new hive. Close the hive; smoke it heavily through the entrance, then narrow the entrance.

It's a good idea to watch the bees near the entrance—if by chance you see any bees fighting, then smoke the hive heavily once again.*

133. Fall inspection of remaining fixed-comb hives

After inspecting the moveable-frame hives in the fashion described above, you'll also need to inspect the apiary's remaining fixed-comb hives. Proceed as described beginning in § 79. If you find hives that

---

* On the subject of uniting hives, it's worth mentioning that it's usually not a good idea to unite colonies in spring. (For more details, see G. de Layens, *Nouvelles expériences pratiques d'apiculture*, ed. Paul Dupont, p. 13.)

CHAPTER 9. FALL OF YEAR ONE

are queenless or too weak, you should unite them with other hives as detailed below in § 204. As we know, since these fixed-comb hives are meant to be transferred into moveable-frame hives the following year, it's best not to harvest any honey from them.

134. Wintering of moveable-frame hives and fixed-comb hives

As the beekeeping season draws to a close, before the first autumn frosts, you should prepare all the hives for wintering, which, as we know, is of primary importance for beekeeping; § 76 describes the wintering of fixed-comb hives.

Let's briefly consider how to prepare moveable-frame hives for wintering as simply as possible.

For successful wintering, we must:
1. Encourage bottom ventilation.*
2. Stop field mice and other predators from getting into the hive.
3. Avoid too much heat loss.

1. You can establish a light flow of air beneath the hive by propping it up by 3/16" (4–5 mm) using two small wedges slid between the hive and the bottom board, from behind (in figure 140, you can see one of these two small wedges, marked w).

Also, to ensure that the bottom board will drain properly, you should also prop it up using two thick wedges, inserted between the bottom board and the stand (figure 140 shows one of these wedges, marked W).

2. To prevent predators from entering the hive, while still allowing air to circulate easily through the bottom front of the hive, you can slide the two metal tabs of the entrances aside and replace them with

---

* You can also winter moveable-frame hives with top ventilation. To do so, remove the slats or the cover boards, and place, across the entire length of the hive, three or four sticks 3/8" (1 cm) thick and cover everything with a tightly packed cushion of moss, for example. Humid air will pass through this cushion and, since you've removed the slats or cover boards, the humidity will constantly escape from the cushion.

two perforated tabs called *mouse guards*; these are metal strips perforated with holes large enough to allow bees to pass through, but too narrow for even the smallest field mice to enter. For example, these holes may be rectangular in shape, 3/8" (7 mm) tall and 1/2" (12 mm) wide.

3. To avoid heat loss, place a straw mat or a cushion of moss atop the cover boards at the top of the hive, or atop the frames themselves.

Now that the fixed-comb hives and moveable-frame hives are ready for the winter, you can leave things as they are until the following spring. Indeed, *it is essential not to disturb the bees during the winter*, since by inspecting them you would risk depriving the colony of the many bees that would be unable to rejoin the cluster due to the cold.

Figure 140. A moveable-frame hive prepared for wintering: *W*, one of two large wedges placed between the stand and the bottom board; *w*, one of the small wedges placed between the hive and the bottom board; *g*, one of the two mouse guards.

Summary

How to harvest honey

When honey season has begun to draw to a close, and if the bees have gathered enough honey, then a beginning beekeeper can practice extracting any surplus honey found in the moveable-frame hives. But no honey should be harvested from the fixed-comb hives, since they're meant to be transferred to moveable-frame hives the following year.

Inspecting hives during autumn

As you're gathering surplus honey from the moveable-frame hives, you should also conduct your fall inspection.

During this inspection you should:
1. See whether each hive contains some brood; if a certain hive has no worker brood, then make a note of the fact that it will need to be united with another hive.
2. Place some honey frames taken from hives with excess honey into hives that are running short, such that each colony has at least 35 lb (16 kg) of honey stored away for the winter.*
3. Remove any surplus honey frames and harvest the honey from them using an extractor.

You should also inspect any remaining fixed-comb hives—if any of them are queenless or excessively weak, you'll need to unite them with other fixed-comb hives.

Feeding and uniting moveable-frame hives

If the season was so poor that no surplus honey is available, or several hives are even running short, then you'll need to supplement these hives' winter reserves with sugar syrup.

Finally, if, due to a particularly bad season, multiple hives have less than 18 lb (8 kg) of honey stocked away, it would be wise to unite them in pairs.

Wintering

As autumn draws to a close, and before cold weather sets in, ready your moveable-frame hives and fixed-comb hives for wintering, then leave all of your hives untouched, without opening them, until the following spring.

---

* As noted previously, more honey will be needed for more northern colonies to survive the winter. See comment to § 125. *Ed.*

Chapter 10
What to Do in The Spring of Year Two

135. End of wintering

When the bees begin actively visiting the first flowers of spring, you can remove all of the wedges used to prop up the hives.

You can also remove all of the mouse guards. For moveable-frame hives, you can replace them with the metal tabs that act as gates ("entrance reducers"); using the gate, completely close the entrance opposite the side of the hive the bees are in, and leave the other gate open to a greater or lesser degree based on the size of each colony.*

136. Inspecting the hives during early spring in year two

As we've already mentioned, you should conduct this inspection on a nice day when the bees are highly active, and when they've already been at work for a week or so. We've seen (§ 79) how to inspect fixed-comb hives; now let's talk about inspecting moveable-frame hives.

The inspection should be done as described in § 118, and you should determine the condition of each moveable-frame hive just as we did for fixed-comb hives in §§ 80–85, but with much greater ease thanks to the moveable frames. In your notebook, record whether each hive

---

* If the hives had top ventilation (see the footnote to § 134), put the cover boards above the frames or the wooden slats in between the frames back into place, just as you removed them before wintering.

is in excellent condition, or weak despite wintering successfully, or strong but wintered poorly, or dead or disorganized.

With moveable-frame hives, since handling the frames is easy and the combs are regularly arranged, the beginner can also easily learn to *judge the quality of the brood*, which is of great practical importance for beekeeping.

137. Judging the quality of the brood

1. If the sealed brood is compact (*B1*, fig. 141) or ring-shaped (*B2*, fig. 141), this indicates that the colony has a good queen, since this pattern is the sign of regular and continuous egg-laying.

2. If the brood is scattered, as shown in fig. 142 (this is quite rare), this generally indicates that the hive has a poor queen,* since her egg-laying is becoming irregular—or that the hive has been struck by a disease called foul brood (§ 284).

3. During this inspection, one can easily see the various possible signs of a disorganized hive, as we've already discussed (§ 84): a hive without brood, and a hive which only has drone brood, whether located in drone cells or worker cells.

138. What should be done with a disorganized moveable-frame hive?

At this time of year, a disorganized (queenless or with a drone-laying queen) moveable-frame hive will generally be very weak, consisting of nothing more than a few handfuls of aging bees.

If you leave the hive untouched, it risks being robbed (§ 92), or having its comb invaded by the wax moth (§ 290). You can liquidate the hive in the following way.

On a nice day, when the bees are highly active, move the hive a certain distance away, then remove, one by one, all of its frames, sweeping the bees onto a board placed on the ground, in the sunlight; unable to

---

* In this case, you should make a notebook entry under this hive's number stating that the hive should be monitored; it may acquire a better queen itself, as you'll be able to tell later if you find brood in a compact or circular formation—if not, you'll need to unite it with another hive at some point during foraging season (see § 132).

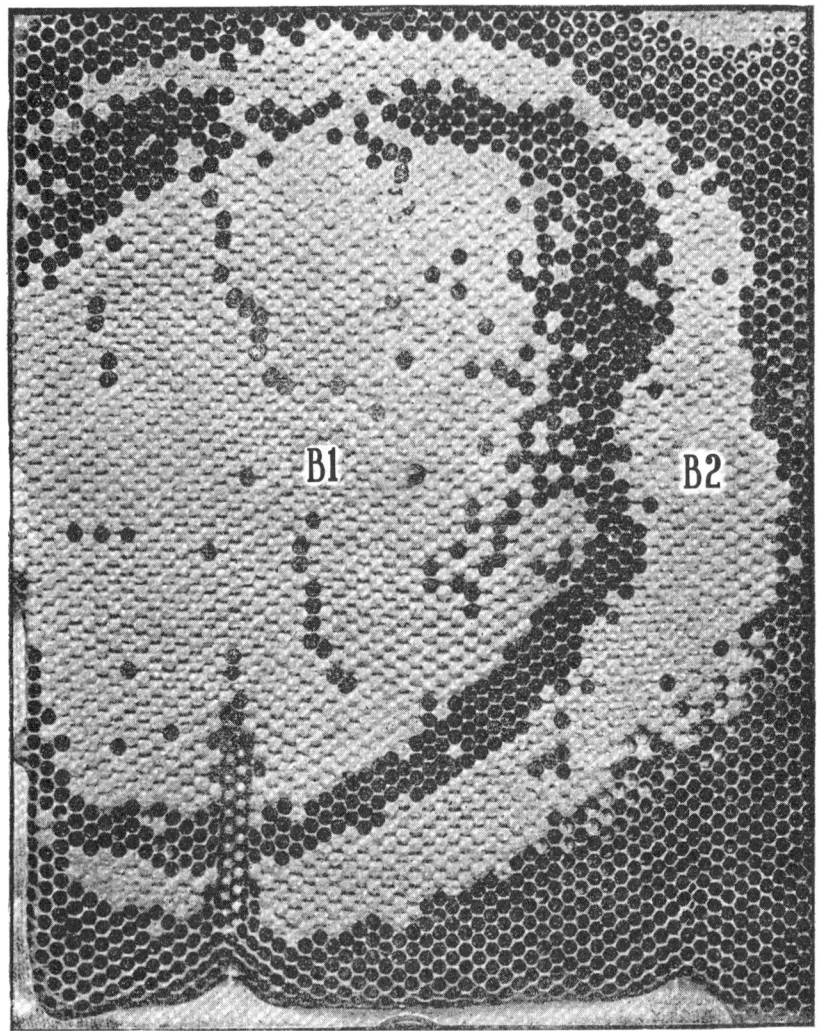

Figure 141. A section of a frame showing the compact brood, *B1*, and a ring of brood cells, *B2*, indicating that the hive has a good queen.

relocate their own hive, the bees will request the hospitality of neighboring hives—a favor willingly granted on a day of nectar flow.

As for the frames in this hive, which may contain some amount of honey, you can move them to the honey room (§ 254).

Alternatively, if the hive is queenless, you can try to give it a queen using any of the methods discussed in § 236.

Figure 142. A section of frame showing scattered brood cells, indicating that the hive has a poor queen.

139. Arranging the frames during the spring inspection in year two

Take advantage of this early spring inspection of your moveable-frame hives to arrange each hive's various frames so as to promote optimal egg-laying in the season to come.

Since during this second year you don't yet have enough built-out frames at your disposal, you'll simply arrange the existing frames in each hive in the following order:

1. Place any frame that contains no brood closest to the wall of the hive that's nearest the open entrance.

2. After this frame, place all the brood frames in the same order you found them in.
3. Now place all the remaining frames, starting with those containing the least honey, which should be placed directly following the brood frames just mentioned.
4. If you have enough frames with foundation, or at least frames primed with pieces of old comb or a bead of wax,* you can easily use them to fill out each hive now—otherwise you'll be forced to add them gradually throughout the season, requiring more monitoring of the hives.

In any case, you shouldn't place foundation frames in between frames that are fully built out but currently empty. Indeed, the bees may excessively lengthen the cells of the already-built comb, while working little on the foundation sheet. This leads to two problems: 1) combs whose cells have been lengthened in this way become too thick to be moved easily; 2) the drawn-out foundation frames will not be thick enough, particularly toward the top.

140. Feeding the hives that are low on honey

If you didn't closely follow the advice on winter reserves the previous autumn, then some of your hives may need feeding in the spring.

For fixed-comb hives, we've already described (§ 88) how to tell whether they need feeding, and how to feed them if they do.

For moveable-frame hives, since you'll have taken notes on each hive during the spring inspection, you'll have a rough idea of how much honey remains in each hive now.

It's been established that, if the spring season yields no nectar, a large hive will need around 22 lb (10 kg) of honey between the spring and the main honey flow—for temperate regions, this means between March and the end of May.

If the first inspection reveals that a certain hive has almost no honey left, should you immediately give it the reserves it's missing, as you did in such cases during the fall?

---

* A starter strip of wax foundation is another option for priming frames. *Ed.*

No, not at all, since some of this food would be of no use if there is a spring flow.

Nor should you feed the bees in multiple, small doses; if the queen believes that a strong spring flow is underway, she might lay too intensely and too early, resulting in various problems caused by so-called speculative feeding (§ 231).

So, in short, here's what you should do.

Give around 4 lb (2 kg) of sugar syrup to each hive that has very little capped honey remaining, either in a bowl (§ 109) or, ideally, by pouring the syrup into the combs (§ 127), or using feeders of various kinds (§ 220).

If you later discover that this feeding wasn't enough, you can add another 2–4 lb (1–2 kg) of syrup to each hive whose stockpiles are low, and continue doing so until the main honey flow is underway.

141. Problems with feeding*

We've just mentioned some of the problems that can result from spring feeding. You will avoid these issues altogether if you leave each hive the amount of honey necessary for survival until the main honey flow begins (§ 125).

Later on, you can avoid feeding problems in spring or fall in another way, when you'll have enough moveable-frame hives to create *a sufficient reserve of capped-honey frames* in your honey room (§ 168).

142. Transferring fixed-comb hives into moveable-frame hives

Now we need to turn any fixed-comb hives remaining in the apiary into moveable-frame hives.

This procedure is referred to as a *transferring*, regardless of how you go about it.

By transferring all of our fixed-comb hives now, we'll have the added advantage of avoiding, as much as possible, creating any natural swarms.

---

* See also § 231.

We've seen that monitoring for and collecting swarms pose difficulties that can be avoided by preventing natural swarming. As one might well imagine, it would be difficult indeed to operate an apiary according to any clear and regular system if each hive could potentially produce multiple swarms every year.

So *knowing how to manage the hives while reducing natural swarming as much as possible* is an important principle in modern beekeeping.

You can transfer hives in two ways, depending on whether the fixed-comb hive is strong or relatively weak—or use other available transfer methods (see § 152).

143. A flipped-hive transfer

A strong hive can be transferred directly into a moveable-frame hive in the way described further below (§ 144), but since this method can prove rather difficult for beginners, especially if the hive is strong, you can use the following very simple method, referred to as a *flipped-hive transfer*.

You should perform this procedure approximately ten to fifteen days before the main honey flow begins—which you can judge roughly by the condition of the vegetation of the region's major nectar plants.

If the transfer comes too soon—for example, in March—then the bees run the risk of excess exposure to any cold spells that may come later.

If the transfer comes too late, the bees may already have begun preparing to swarm, and you'll have to collect the resulting swarms and return them to the hive.

Start by smoking the fixed-comb hive you're about to transfer, then move it several steps away from its present location, along with its stand and bottom board. Then, where the stand previously stood, dig a hole in the ground big enough to hold approximately half of the fixed-comb hive if it were turned upside down.

After smoking the relocated fixed-comb hive intensely once again, turn it upside down so that you can place the top of the hive in the bottom of the hole in the ground.

You should have prepared in advance a board with a square opening that is a bit narrower than the diameter of the fixed-comb hive.

After flipping the fixed-comb hive as described (*F*, fig. 143), place the opening of the board atop the opening of the flipped hive, and—using bricks, for example—support the edges of the board on the side opposite the hive to ensure that it is positioned horizontally; then, atop this, place (*M*, fig. 143) a moveable-frame hive with ten or so frames with foundation, or, barring that, amply primed with pieces of old comb. Make sure, of course, that the bulk of the frames are positioned above the opening in the board. Open the entrance of the moveable-frame hive that is on the same side as the flipped fixed-comb hive, leaving the other entrance shut.

Next, in one way or another (with some rags, some cow manure, etc.), fill in the gap between the flipped fixed-comb hive and the bottom of the bottom board. Now the bees will only be able to pass through the entrance of the moveable-frame hive.

If the flipped hive is strong enough, and

Figure 143. Flipped-hive transfer. *M*, moveable-frame hive placed atop a fixed-comb hive, *F*, turned upside-down.

if the season is abundant enough in nectar, then the bees will have climbed into the moveable-frame hive by autumn, and installed themselves naturally in their new home, which you can easily confirm by the presence of brood in the frames—this means that the flipped-hive transfer was a success.

In this case, all you need to do is place a stand supporting an ordinary bottom board in front of the hive; then, after smoking the hive, move it from the board with the opening to the new stand. Then remove the fixed-comb hive from its hole in the ground, fill in the hole with some dirt, and place the moveable-frame hive, along with its stand, directly above the refilled hole in the ground, making sure it is completely level.

You can make use of the comb from the dissolved hive as described above in §§ 85 and 86.

If a poor foraging season has prevented the bees from climbing into the moveable-frame hive and installing themselves there completely, then in autumn you'll need to remove the moveable-frame hive and restore the fixed-comb hive to its normal position, readying it for wintering.

144. Direct transfer

The transfer method described above often works with strong hives, but almost never works with relatively weak ones. So the latter will need to be handled differently.

The quickest method is a *direct transfer*—but this method is quite difficult to pull off. To ensure success, a beginner is advised to ask for help from an experienced beekeeper.*

Direct transfer involves removing all of the comb containing worker brood and empty worker cells or cells full of honey from the fixed-comb hive you're going to transfer, cutting them up and arranging them in the frames, where they're held in using strings, and, finally, placing the frames filled in this manner into a moveable-frame hive where one then installs the bees from the fixed-comb hive.

You should perform this procedure during April (or during March in the south of France). If it's done too early, the colony may, due to periods of prolonged cold, be unable to adapt in its new home—but if it's done too late, the transfer may be complicated by the large quantity of new, uncapped honey in the hives.

Here's how to complete a direct transfer.

145. Preparing frames to hold comb from the fixed-comb hive

In order to attach pieces of comb to moveable frames later, hammer some upholstery tacks halfway into the frame's end bars at the points marked *H, G, F, E, D, C, B, A* (fig. 144).

---

* See § 230 for other transfer methods.

Wrap one end of a string around the tack marked *H*, then hammer it into the frame to fasten the string; then, wrap the string around tack *G* and hammer it into the frame; then *F*, then *E*, and so on until *A*. When you reach this final tack, leave a length of string hanging (fig. 144) equal to the length used up thus far. Turn the frame over and hammer more tacks halfway into the frame at the same points, just as before, but without wrapping the string around them for now.

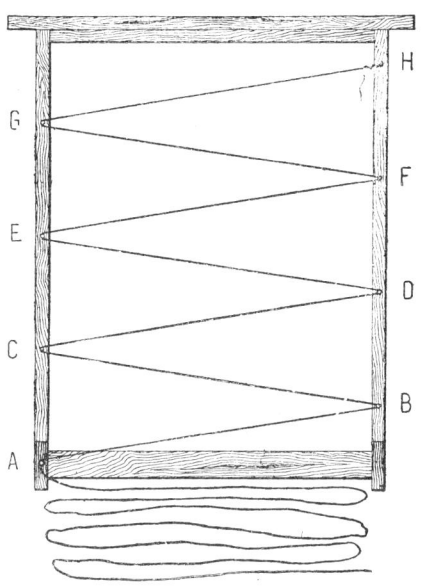

Figure 144. Preparing a frame for a direct transfer.

Have five or six frames, prepared in this fashion, in your honey room, ready for the first hive transfer.

146. Driving bees from a fixed-comb hive you're going to transfer*

This procedure should only be performed when it's warm enough (60–70°F or 15–20°C), and when the bees are highly active.

Let's begin by lightly smoking the fixed-comb hive through the entrance; detach it from its bottom board using a chisel (hive tool), then prop it up using a wedge and smoke some more underneath the hive.

---

* For this procedure, you'll need: 1) a sheet or piece of cloth; 2) a piece of black fabric that is at least as large as the hive; 3) two empty fixed-comb hives with no comb inside, roughly identical to the hive you're planning to transfer; 4) a fixed-comb hive super or a stool; 5) a smoker; 6) a veil; 7) two sticks of wood 12–16" (30–40 cm) in length and 3/8" to 3/4" (1–2 cm) thick; 8) several pieces of thick wire bent at a right angle at two ends; 9) a chisel or a knife; 10) a wedge. It would be ideal to have an assistant for this procedure.

When you're done smoking, turn the hive upside down and move it into the shade some distance away, placing it sturdily atop an upside-down stool or a super. On the spot where the hive previously stood, place an empty fixed-comb hive to welcome the bees who return from foraging.

Smoke the upside-down hive lightly as necessary to calm the bees; meanwhile, place another empty fixed-comb hive atop the first hive,

Figure 145. A beekeeper driving the bees of one hive into the basket of an empty fixed-comb hive.

attaching them using wire staples, such that the second hive touches the upside-down hive on the side of its entrance, but is slightly elevated on the other side, creating a gap between the two where you'll be able to observe the bees passing from one hive to the other. Now, stop smoking, and strike both sides of the upside-down hive with a stick (fig. 145).

Begin with a series of rapid but moderate blows with the stick near the bottom of the hive, for more than five minutes; then, continue drumming with the stick, making your way upward along the surface of the hive. This is called *drumming*.

Agitated by this constant noise, the bees will stuff themselves with honey and finally make up their minds to climb, in large numbers, into the empty hive above them; this is when we can watch them in the gap we left between the two hives (fig. 145). If we have enough experience with bees, we'll even be able to spot when the queen moves into the upper hive, since she'll usually walk atop the crowd of bees as they themselves climb into the hive.

Figure 146. A beekeeper examining the piece of black fabric that was placed beneath the basket full of bees, in search of eggs that may have fallen there.

Regardless of whether or not you spot the queen, when it seems that the majority of the buzzing group of bees has made its way into the empty hive, unfasten the wire staples and place the empty hive, along with the bees inside, on some black fabric spread atop the sheet on the ground, in the shade, propping up the hive using a small wooden wedge.

A certain number of bees will still remain behind in the upside-down hive. Now, pick up the hive to return it to its natural position, then shake it by striking its edge lightly against the sheet, not far from the new hive that now contains the majority of the bees. We can shake it several times in this fashion, in various places; the bees who fall out of it will naturally rejoin the others in the new hive placed atop the black fabric.

If, after a half-hour has passed, the bees still remain calmly in the empty hive, you can be almost certain that the queen is there with them. We can make sure by lifting the hive now containing the bees and inspecting the black fabric placed beneath it. If, among other small bits of debris, we see some eggs that have fallen onto the fabric (fig. 146)—their white color will make them clearly visible against the fabric despite their small size—we have proof of the queen's presence.

If you find no eggs, refer to § 150.

147. Removing the comb from the fixed-comb hive and placing it in the prepared frames*

Move the hive containing the comb to a table in your honey room. Use pliers to remove the sticks that were supporting the combs,** then use a hooked knife to carefully detach, one by one, all of the comb from the fixed-comb hive and place them on a table. If some are still occupied by bees, sweep them off outside next to hive on the black fabric. Some bees may also be left on the bottom of the hive you've just removed the comb from—shake them out onto the black fabric as well.

Now, lay an ordinary frame, containing no comb at all, flat on the table. Taking each comb removed from the fixed-comb hive, cut it into several pieces and arrange them in a temporary frame, such that the pieces all touch one another and fill the frame completely. Take care to place the brood comb near the middle of the frame, with all of the

---

* For this procedure you'll need: 1) a bowl filled with water for washing your hands or tools from time to time; 2) a knife bent at a right angle (fig. 96); 3) pliers.

** If the fixed-comb hive is wooden, remove one wall so as to take out the combs more easily. If it's a straw hive whose outer casing is of no use, you can saw it lengthwise or cut it with clippers to make the procedure much simpler.

pieces of brood touching each other. Then, surrounding the brood, and especially toward the bottom, fill in the rest of the frame with pieces of empty worker comb (this is the arrangement seen in figure 147).

During this procedure, cut out all the sections of comb containing drone cells, regardless of whether they're empty or contain honey or drone brood. Leave them on the table with all the small pieces of comb that are unusable.

So the bits of comb have now been positioned precisely in the temporary frame, and we've been sure to trim every piece on all sides with a knife, since this will allow the bees to weld them together more quickly. Now, take a frame prepared with string as described above (fig. 144), and place it on the table so that the side with the zigzagging string is facing the bottom, then transfer all of the pieces arranged in the temporary frame into the permanent frame with the string, keeping them in exactly the same position. Next, take the part of the string that hasn't yet been fastened to the tacks (that is, the section of string shown at the bottom in figure 144), and stretch it across the upper face of the frame, wrapping it around the tacks that have only been hammered half-way into the frame (§ 145), stretching it firmly as you hammer each consecutive tack all the way in.

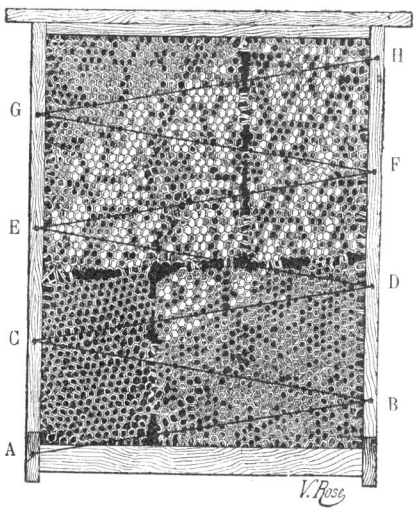

Figure 147. A frame filled with pieces of comb taken from the hive destined for transfer, ready to be placed into the moveable-frame hive.

Now the frame is ready to be placed as-is into the moveable-frame hive (fig. 147).

Lift the finished frame carefully upright and straighten out any pieces of comb that may have been jarred out of line, then place it in the frame box.

Let's assume we've salvaged enough pieces of comb from the fixed-comb hive to form three frames that contain brood and have been filled out with empty comb. Using the remaining comb, we can use the same procedure to prepare frames containing honey and the remaining empty worker cells—but be sure to place any pieces of comb with honey near the top of the frames.

Assuming that we've had enough comb to prepare three brood frames and three frames containing some amount of honey, we can now arrange them in the moveable-frame hive, still in the honey room, in the following order:

1. One frame with honey alongside the hive wall.
2. The three frames with brood.
3. The other two frames with honey.
4. Three or four frames primed with foundation, or at least with bits of comb.

148. Moving the bees to their new hive

With the moveable-frame hive ready and waiting in the honey room on a sheet and propped up on one side with a wedge, gently move the hive from the black fabric, containing the bees, alongside the new moveable-frame hive—then, with a sudden and forceful blow, shake the bees out in front of the moveable-frame hive just as you would when transferring a swarm into a moveable-frame hive (§ 107).

Once the bees have climbed into their new home, leave the newly populated hive in the honey room until an hour or two before nightfall.

Then, take the empty hive that took the place of the fixed-comb hive whose bees you transferred, and put it on the ground. Now, in its place, gently set onto the stand the moveable-frame hive that now contains the transferred bees.

Open the hive entrance that is on the same side as the frames, just wide enough for two or three bees to pass at a time, in order to avoid any robbing that might be provoked by the strong smell of honey that the transferred hive emits. Any bees who, back from the fields, are now in the hive that was placed temporarily in the location now occupied by the moveable-frame hive, will eventually make their way into their new home of their own accord—but it's best to go ahead and shake them out in front of the moveable-frame hive.

The day after next, you can widen the entrance.

149. What to do with the comb that wasn't used during the transfer

All of the bits of comb with drone cells containing honey, as well as all the other bits of comb with honey, should be placed in any order in between two screens attached to the two faces of a frame as shown in figure 148; to ensure that all the honey is removed quickly, be sure to

Figure 148. Bits of comb not used during the transfer, placed in between two screens and placed into the transferred hive to allow the bees to remove the honey.

use a knife to uncap any capped honey cells included in these pieces of comb.

A few days later, near evening, place these screen-covered frames in the newly transferred hive, in the spaces following the empty frames. The bees will remove the honey and transfer it to other cells.

Later, when no honey is left, remove the screened frames from the hive and take out all of the pieces of comb. Save any combs containing worker cells, since they can be used later to prime empty frames.

As for the bits of comb containing drone cells, whether empty or with drone brood, as well as any scraps of worker brood, you can set them aside to be melted down as soon as possible (§ 277).

150. What if there aren't any eggs on the black fabric during the transfer?

If you don't find any eggs on the black fabric (fig. 146), it's highly likely that the queen isn't present with the other bees in the empty hive you transferred them into. When this is the case, the bees are often agitated, and some of them may go back, along with the bees returning from the fields, into the empty hive placed there temporarily.

You'll need to look for the queen, who has probably remained behind on a comb inside the hive slated for disassembly. If you find her there, take her gently in your hand—she won't sting when handled—and place her underneath the hive on the black fabric.

If you can't find the queen, then she is likely at the bottom of the hive or, in rare cases, has fallen to the ground somewhere. In the former case, don't cut out the comb until you've carefully sawed open the hive. In the latter, as you scan the ground you may notice a small, isolated group of bees—and it's in this group that you'll find the queen.

Finally, if you've failed to find the queen in any of these places, or if she has been killed during the transfer procedure, then continue the transfer as if the queen were there; with brood of all ages at their disposal, the bees will produce a queen, although this will represent a significant setback for the colony's development.

Throughout the procedure described above, be sure *not to discard any bit of comb containing honey outside the hive*, since this could provoke robbing.

151. Monitoring the transferred hive

A few days later, near evening, it's a good idea to open the transferred hive to see whether any bits of comb need to be straightened up. As for the strings, the bees will take it upon themselves to chew through them bit by bit and ultimately discard what remains of them through the hive entrance.

152. Problems with direct transfer. Other transfer methods

If applied to all the fixed-comb hives, the direct transfer method we've just described will immediately turn all of your fixed-comb hives into moveable-frame hives—but there's no denying the fact that direct transfer may present serious problems for a beginner working without the advice of a more experienced beekeeper. So if as a beginner you have no one to turn to for help and are hesitant to risk a direct transfer, you may choose one of the following alternative methods:
1. You can collect natural swarms from fixed-comb hives and install them in a moveable-frame hive as described in § 107.
2. You can transfer fixed-comb hives into moveable-frame hives using the superposition method (§ 230, point 1).
3. Leaving aside any hives that have become too weak, which should be kept for the time being in fixed-comb hives, you can use any others to produce artificial swarms, a procedure described below (§ 230, point 2). This final alternative method is the best.

Summary

Spring inspection
When wintering is over, remove the mouse guards and the wedges from the hives and put the metal gates at the entrances of your moveable frame hives.

Inspect all of your hives in early spring, both fixed-comb hives and moveable-frame hives; the latter are much simpler to inspect, and you'll be able to easily judge the quality of the brood, which will allow you to assess the hive's condition. Use this opportunity to arrange the frames inside your moveable-frame hives in a way that will best promote egg-laying.

Spring feeding

If you've carefully followed the advice given in the previous chapter, you shouldn't need to feed your hives in the spring.

Transferring

Now is the time to turn any remaining fixed-comb hives into moveable-frame hives—that is, to complete a *transfer*. There are five ways of doing so:

1. *Flipped-hive transfer* (§ 143). Place the moveable-frame hive atop a fixed-comb hive that has been turned upside-down. This method is simple, but not always successful. It is more likely to succeed with a strong hive and in a good year.

2. *Superposition transfer* (§ 230, point 1). Place the fixed-comb hive atop the moveable-frame hive. This method is simple, but not always successful. It is more likely to succeed if the fixed-comb hive is rather small, and in a year with an abundant honey harvest.

3. *Natural-swarm transfer* (§ 107). Install a natural swarm in a moveable-frame hive. This method is simple, but assumes that swarms have emerged—and watching for and collecting swarms can be a lot of trouble.

4. *Artificial-swarm transfer* (§ 230, point 2). Use strong fixed-comb hives to produce an artificial swarm. An excellent method, but doesn't allow you to transfer all of your fixed-comb hives simultaneously.

5. *Direct transfer* (§ 144). Place all of the fixed-comb hive's contents into a moveable-frame hive. A quick and efficient method that allows you to turn all of your fixed-comb hives into moveable-frame hives. Difficult for a beginner with no help from an experienced beekeeper.

Chapter 11

What to Do in the Summer And Fall of Year Two

153. Monitoring hives during honey season

During the honey season of year two, a beginner should gain experience with general monitoring of the apiary and inspecting moveable-frame hives.

Based on everything we've assumed thus far, you should now have:
1. Multiple moveable-frame hives that have survived the winter, and that were populated the previous year with swarms.
2. Strong hives currently undergoing flipped-hive transfers.
3. Some weaker hives that have been transferred directly into moveable-frame hives.

Here's how you should monitor these three types of hive.

1. If the hives in the first group were filled completely with frames primed with foundation or bits of comb as described above, then they'll usually require no further attention until it's time to harvest their honey. However, a beginner should inspect them from time to time just to make sure everything's in order.

2. For hives currently undergoing a flipped-hive transfer, you should check to see that the connection between the two hives remains well sealed, and whether all of the bees are entering and exiting through the entrance of the moveable-frame hive.

It may be interesting to glance from time to time at the frames in the moveable-frame hives placed atop the flipped hives to see whether the bees have begun to build comb there, or have already installed themselves there.

If the bees have installed themselves quickly in the moveable-frame hive, you'll need to add some more frames primed with foundation or comb, as necessary, in addition to those already in place.

3. As for the hives that were transferred directly, we've already mentioned (in § 151) that they'll need special care—whether adjusting parts of comb that may have been jostled out of place between the strings, or adding new frames primed with foundation or bits of comb.

The most vexing discovery one can make in a directly transferred hive is that the pieces of comb were placed too loosely atop one another between the strings, and have collapsed.

If this is the case, then visit the hive around late afternoon and rearrange the poorly positioned pieces of comb inside the frames. Remove any frames that you're unable to rearrange, and put everything back into place.

154. Strengthening a weak hive

As you monitor your apiary, you should always have a rough idea of the relative strength of each colony, even without having to actually inspect the hives.

If during the honey flow you notice that a colony is growing weak or has little honey, but still has worker brood, you can strengthen it as follows.

The simplest way to strengthen the hive is to swap its location with that of a prospering hive.

You should carry out this procedure in the morning—between nine and eleven, for example, on a day when the bees are highly active.

Smoke the strong hive through its entrance, then smoke the weak hive in the same way. Place the latter on the ground, then bring the strong hive and place it on the stand previously occupied by the weak one. The bottom boards and stands will remain in place; only the two hives will have switched locations. Smoke both hives once more through the entrances. Now the procedure is complete.

The bees returning from the fields to the strong hive's previous location will recognize their old stand, enter the weak hive you've placed there. They may leave again when they don't recognize the hive as their own, but in the end they'll return once and for all.

Now, the weak hive that you've relocated will see a considerable population increase and may eventually grow into a flourishing hive.

Meanwhile, the strong hive that now occupies the previous location of the weak one will show very low activity levels, since it will have lost a large percentage of its population, while only gaining those few bees from the weak hive who happened to be in the fields when the hives were swapped. But don't forget that this hive has a lot of brood, and more than enough bees to tend to it. So don't be surprised if, after some time has passed, and once the brood has emerged, you discover that this hive has become highly active again.

This very simple procedure will thus result in two solid hives in place of one strong one and one weak one.

155. What to do when the combs have collapsed

You may on occasion find combs that have collapsed inside a hive, in situations unrelated to that mentioned in § 153. If this is the case, the collapsed comb may have contained honey that will now flow onto the bottom of the hive, leak outside, and risk attracting robbers. In this case, you should immediately move the hive into your honey room, along with its bottom board. Put another bottom board in its place, with an empty hive on it, to welcome any bees returning from the fields.

In the honey room, remove all of the collapsed comb, putting any that was left undamaged back in its former position; add some new foundation and, after cleaning the bottom board, return the hive to its original location.

156. What causes comb collapse

There are two primary causes of comb collapse:
1. The foundation you used was made with artificial wax, or real wax that was too thin;
2. The hives were exposed to full sun, and the summer temperatures were so high that they caused the wax to soften.

Here's how this problem can be avoided:

1. Never buy foundation that wasn't made of pure beeswax. *Buying low-priced foundation of dubious purity is never a good deal.*

You'll learn later (§ 280) how to make foundation yourself; and those who have to purchase it will learn (in § 280) a very simple way to tell whether or not it is genuine.

2. Earlier, we suggested keeping your hives in the shade as much as possible. If this is impossible, it's a good idea to place some straw mats on the roofs of your hives during the summer, secured in some way to prevent the wind from blowing them away.

157. Swarm prevention

Colonies fully installed in large moveable-frame hives will rarely produce natural swarms, with the exception of certain bee races. If you do have swarms, however, return them to their parent hives as described in § 113, which discusses afterswarms.

158. Supersedure (natural queen replacement)

Beekeepers have long practiced artificial requeening—that is, replacing queens* they believe to be too old with young ones bred especially for this purpose. Now, it's an established fact that it is *preferable to let the bees supersede (replace) their queens themselves* when their fertility ebbs—whenever the bees see fit to do so.**

159. Fall inspection, honey harvest, and wintering

During the fall of the second year, you can carry out the fall inspection and honey harvest simultaneously as described in § 124, always being sure to leave enough honey for wintering in each hive.

Any extracted frames should be returned to the hives in the evening, all on the same day, to allow the bees to clean them. To avoid

---

* See § 237 and the following.

** For more details, see G. de Layens, *Conseils aux apiculteurs*, p. 19 (ed. Paul Dupont).

robbing, be sure to narrow the entrances of all the hives for several days, so that only two or three bees can pass through at a time. There is no real problem with leaving these frames in a hive throughout the winter, as long as it is well ventilated.

If the year has been a particularly bad one, beginning beekeepers may not have enough honey frames at their disposal to give each hive sufficient honey reserves; in this case, you'll need to feed some of the hives as described in § 127.

If some hives being transferred using the flipped-hive method have not fully installed in their moveable-frame hives—you can tell by the absence of brood in the frames—you'll have to put the fixed-comb hives back into their old position, remove the moveable-frame hive, and postpone the transfer until next year.

Finally, all the hives should be prepared for wintering (see § 134).

Summary

Monitoring your hives during honey season
A beginner should monitor all of the hives during the honey season, adding new frames primed with foundation or pieces of comb as needed.

If a certain hive has grown weak, you can strengthen it by swapping its location with that of a strong hive. If some hives have comb that has collapsed, whether they are combs used during a direct transfer, or combs built on cheap foundation made with artificial wax, or even combs that were softened by excessive heat, you should remove all of the damaged parts and put the hive back in order.

If some of your moveable-frame hives produce swarms, return them to their hives of origin.

Supersedure (natural queen replacement)
Don't worry about replacing queens in your hives—leave it to the bees to do so naturally.

Finally, complete your fall inspection and honey harvest, and prepare your hives for wintering.

Chapter 12

What to Do in Year Three

160. The end of wintering. Year three

If all of the previous year's transfers were a success, then by the start of year three your apiary should contain moveable-frame hives only.

If this is not the case, you should handle your fixed-comb hives just as you did in the first year or transfer them as described beginning in § 142.

But, for simplicity's sake, let's assume in this chapter that all of your colonies are now housed in moveable-frame hives—some for two years now, and others transferred just last year.

When winter ends, a beekeeper should remove the mouse guards and wedges, and position the metal tabs on the hive entrances as you did at the start of year two.

When the bees have been working actively for about a week, inspect all of the hives. If you have enough frames full of empty comb (with worker cells) in your hives or in your honey room, you can arrange them as follows.

161. Arranging your frames in the spring

Arranging your frames in the spring is important for ensuring that the colony functions normally during the season to come.

This arrangement will:
1. Give the queen enough room to lay her eggs.
2. Prevent the brood from expanding across too many frames, which can sometimes cause problems during the honey harvest.

3. Allow the bees to draw out a certain number of frames, which can be very useful.*

After smoking a hive, inspect it as described in § 118, then arrange the frames in the hive as follows.

Place worker-cell frames (with no honey) to the right and left of the brood frames, which generally contain honey near the top, to encourage the queen to lay eggs in this part of the hive only. Of course, the brood is always on the side facing the open entrance.

Then, at the other end of the hive, place the frames containing the most honey,** alternating them with empty frames that have only been primed at the top. Figure 149 shows a lengthwise cross-section of a hive showing this frame arrangement. Figure 150 depicts one of these frames, containing brood and honey, cut in half and shown at an angle.

Any hives for which you don't have enough frames, and which can't be arranged in this way, should be handled as described in § 139.

As you can see, this arrangement leaves the queen plenty of room to lay her eggs. In the meantime, the bees can store their honey without interfering excessively with the egg-laying process—and they'll also have some frames to build comb in.

---

* The question of how much honey bees consume to produce wax is a long-standing question. Recent experiments by de Layens in France and Viallon in the United States have shown that bees require around 6 pounds of honey to produce 1 pound of wax. But this figure may vary depending on circumstances.

In any case, the question of whether a bee needs this much honey to produce wax, from a purely physiological standpoint, is of little practical concern for a beekeeper. A beekeeper needs to know whether a colony that is allowed to build a certain amount of comb sees any kind of productivity boost from such timely comb construction. Experiments have shown that a colony that is allowed to build several combs gathers at least as much nectar as a colony that is not. See G. de Layens, *Nouvelles expériences pratiques d'Apiculture*, pp. 1–12, ed. P. Dupont. See also, by the same author, *Conseils aux Apicultueurs*, as well as the experiments of Abbot Martin, President of the Eastern Beekeeping Society (Société d'Apiculture de l'Est) in "Faut-il faire construire de la cire aux abeilles?" (in *Apiculteur*, 1892, p. 448).

** It's a good idea to use a knife to uncap honey that is near the top of combs containing brood, which can promote the progress of egg-laying and make the brood area more compact, since the bees will remove all of this honey and replace it with brood.

So the three requirements we mentioned earlier have been met.

Moreover, if you've taken these steps, and make sure not to leave the entrance open any wider than 3–5" or 8–12 cm (depending on the size of the colony) throughout the foraging season, then you won't have to worry about the brood spreading too much across the combs. In fact, if during the peak honey flow, your entrances are too wide and your hive is propped up on wedges, allowing fresh air to reach too many frames at a time, then the queen will tend to extend her egg-laying across too many frames (see the end of § 33).

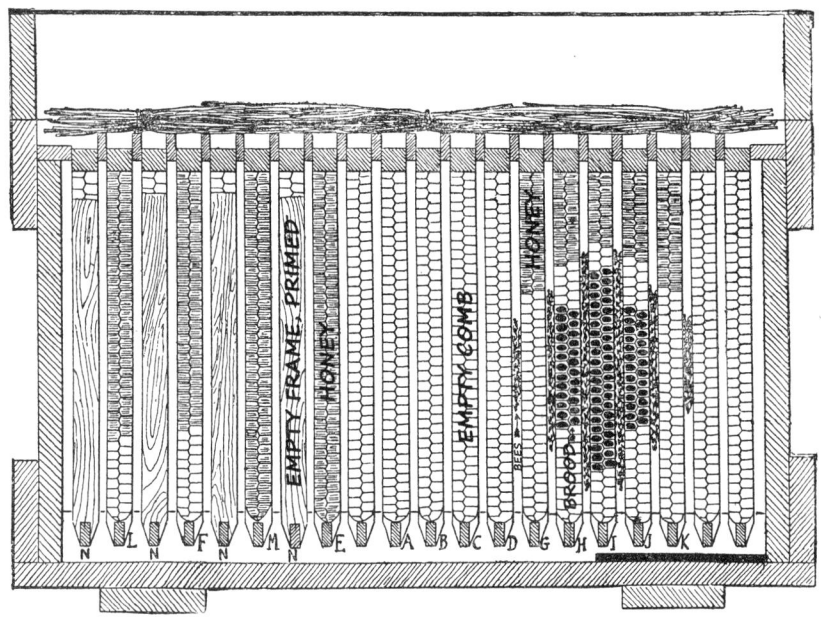

Figure 149. Frame arrangement in spring (lengthwise cross-section of the hive): *H, I, J* are frames with brood and honey; *G, K* are frames with honey near the top; *L, F, M, E* are frames with honey; *N, N, N, N* are frames primed with bits of comb; the remaining frames have been drawn out but are still empty of honey.

162. Maintaining or increasing the number of hives

Since it's best to avoid natural swarming, and since one often has some queenless hives on hand that need to be dissolved, or weak hives to be united with other colonies, you'll need to have some new colonies available to replace those that are lost. Also, you may want to increase the number of hives in your apiary.

You can increase your number of colonies in two major ways:
1. By artificial swarming (§ 163).
2. By purchasing new hives from another region* (§ 164).

163. Artificial swarming**

Artificial swarming is a procedure by which a beekeeper extracts a swarm himself and installs it directly into the hive it's meant to occupy.

There are multiple artificial swarming methods; for now, let's just describe one of the best—one that isn't too hard to carry out.

Ideally, you should conduct an artificial swarming procedure around two weeks prior to the main honey flow, on a nice morning between nine o'clock and noon, when the bees are highly active.

We must mention that when the spring has been highly unfavorable for foraging and for colony development, it's best not to produce an artificial swarm, unless a strong autumn honey flow is all but assured.

The procedure we're about to describe calls for producing an artificial swarm—but

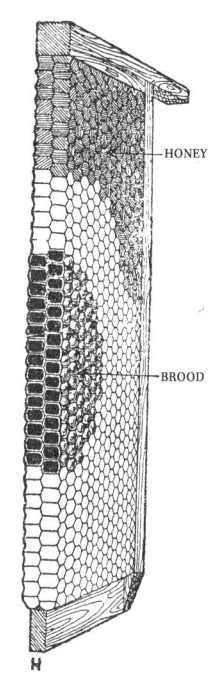

Figure 150. Lengthwise cross-section of a frame. It corresponds to frame *H* in fig. 149.

---

* Three other options include buying nucleus colonies (nucs), buying package bees, or attracting swarms to swarm traps (bait hives). If you have a thriving wild bee population where you live, catching swarms is by far the best way to get quality bees that would be locally adapted and disease-resistant. If you have to buy bees, give preference to local breeders who breed from local or locally adapted hygienic stock. Obtaining local disease-resistant bees is of crucial importance for sustainable treatment-free beekeeping. *Ed.*

** See also § 234.

not with a single hive. Rather, *two strong hives* will contribute to this swarm.

This system ensures that you'll end up with three hives that, altogether, often contain more honey and wax at the season's end than the two starting hives would have.*

Let's say that we've chosen two strong colonies in our apiary, which a preliminary inspection has shown to have plenty of brood and bees.

Let's take hives A and B as shown in figure 151, for example. Assume that we've placed a third hive, C, nearby, containing neither frames nor bees. After smoking hive B, remove all of the frames with bees inside it one by one, sweeping the bees off the comb and back into hive B; then, take the frames (now brushed free of bees) and place them one by one into hive C, in the same order you found them in. Each time you insert a frame, be sure to cover hive C again to protect it from robbing. In place of the frames removed from hive B, insert some new ones, alternating between frames primed with foundation and with bits of comb.

It's a good idea to leave one frame containing both brood of all ages and honey in hive B. Place it in the next-to-last position on the same side as the open entrance.

So hive C now contains all of the frames previously in hive B, except for the one frame you left there.

Hive C should now be placed in the previous location of the other strong hive, A, which, in turn, we move to a new position, quite far away, on a new bottom board atop a stand (fig. 152).

What will happen now?

The bees in hive A that were away foraging will return to their usual location, recognize their old stand, and enter the new hive C—where they'll find a large amount of brood and go about caring for it. When they discover that the queen is no longer in the colony, they'll be slightly agitated at first, but some time later they'll decide to build queen cells, taking advantage of the young brood at their disposal.

---

* See "Expériences sur l'essaimage artificiel" by G. de Layens (in *Apiculteur*, 1894, p. 403; and 1895, p. 51). *Note from Mr. Gaston Bonnier:* "I've been present during de Layens' experiments, and have obtained the same results when repeating them at my own apiary."

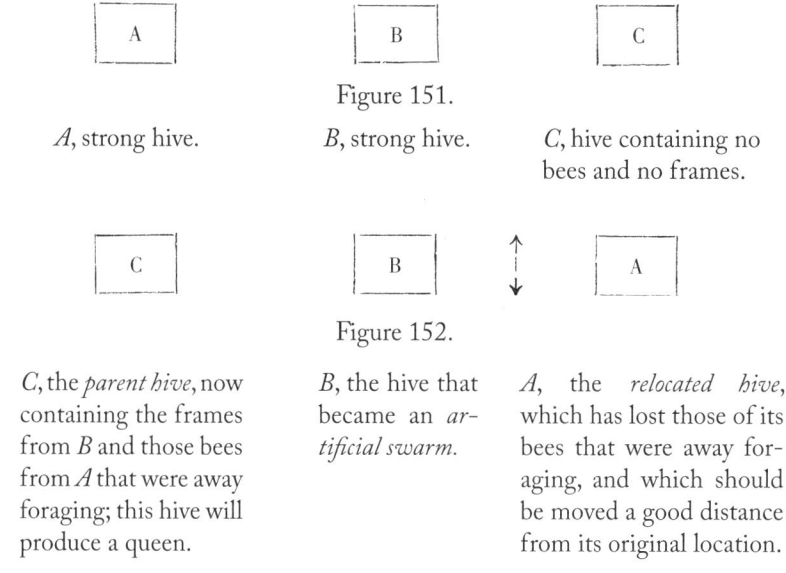

Figure 151.

A, strong hive. B, strong hive. C, hive containing no bees and no frames.

Figure 152.

C, the *parent hive*, now containing the frames from B and those bees from A that were away foraging; this hive will produce a queen.

B, the hive that became an *artificial swarm*.

A, the *relocated hive*, which has lost those of its bees that were away foraging, and which should be moved a good distance from its original location.

Let's look now at hive *A*—that is, the strong hive that was relocated. This hive has kept all of its frames, along with all of the bees on them, and has lost those bees that were out foraging and have now returned to hive *C*. Over the next few days, you'll also notice some bees leaving—to rejoin hive *C*, in all likelihood. But at a certain point, as the brood begins to emerge, you'll notice that hive *A* is beginning to bustle. This relocated hive *A* is very unlikely to swarm.

What about hive *B*, which now houses the artificial swarm? Its bees will begin making arrangements in their new home, and the rapid coming-and-going you'll notice around the hive entrance will indicate that their activity level is reaching a fevered pitch.

So, in sum, we now have three hives in place of two:

1. Hive *C*, with plenty of brood, which will produce a new queen. It contains the comb from hive *B* and the foraging bees from hive *A*. This is the *parent hive*.
2. Hive *A*, with plenty of brood and with a queen. Though it has lost its foragers, its population will soon rebound. This is the *relocated hive*.
3. Hive *B*, containing a queen and all of the bees of the parent hive, who will build new comb, in which new brood will be housed. This is the *artificial swarm*.

Returning an afterswarm from hive C, and monitoring this hive. If, contrary to expectation, hive *C* produces an afterswarm—an event that is always presaged by the queen's song—then it will emerge 13 or 14 days following this procedure. Collect it in a fixed-comb hive and place it in a cellar until evening on the following day—then, put it back into hive *C* by shaking the bees out in front of the entrance.*

If you'd like to make certain that hive *C* has indeed produced a new queen, and that the new queen is fertile, you can check for sealed brood forty-five days later (in general, you can even find some before this date). If, by chance, there is still no brood by this date, it means that the procedure failed, and you'll have to unite this hive with another (§ 132).

164. Hives purchased in another region

Experience has shown that one shouldn't maintain or expand an apiary indefinitely using artificial swarms alone.

If the population of bees in the apiary breeds exclusively with itself, then in the long run it could degenerate, producing colonies with lower activity levels. In any case, then, it's recommended to buy a few colonies, from time to time, from a region distinct from the one you live in; introducing new hives will maintain and even boost the overall activity level in your apiary.

These hives should be purchased under the conditions described above, beginning in § 65.

165. General monitoring of your apiary during year three

As described above, during the spring inspection you should have made entries in your notebook regarding the condition of each of your hives.

Once your apiary contains a certain number of hives, you may find it useful to make a table, where you'll record the condition of the hives and any procedures you carry out under successive dates.

For example, look at the following table, made for 20 colonies.

---

* See also § 233.

CHAPTER 12. YEAR THREE

Table showing apiary progress during one season

Hive No.	March 25		Notes	May 20 Artificial swarming	Apiary after swarming	August 25			
	Frames brood	Honey in hives				Honey before harvest		Honey harvested	
		lb	kg				lb	kg	lb	kg
1	5	9	4	Added 11 lb (5 kg) honey from No. 6	No. 21. Swarm taking the place of parent hive (No. 1) swapped with relocated No. 3	No. 21. Art. swarm	49	22	13	6
2	4	22	10			No. 2	51	23	15	7
3	4	13	6			No. 1. Parent hive	35	16	0	0
4	3	13	6			No. 4	68	31	33	15
5	6	9	4	Added 11 lb (5 kg) honey from No. 12	No. 22. Swarm taking the place of parent hive (No. 5) swapped with relocated No. 7	No. 22. Art. swarm	37	17	0	0
6	1	15	7	Brood scattered, colony dissolved						
7	4	13	6			No. 5. Parent hive	73	33	37	17
8	3	15	7			No. 8	51	23	15	7
9	4	13	6		No. 23. Swarm taking the place of parent hive (No. 9) swapped with relocated No. 11	No. 23. Art. swarm	73	33	35	16
10	4	13	6			No. 10	62	28	24	11
11	6	7	3	Added 11 lb (5 kg) honey from No. 12		No. 9. Parent hive	68	31	31	14
12	0	22	10	Queenless, colony dissolved						
13	4	13	6			No. 13	99	45	62	28
14	5	13	6			No. 14	88	40	51	23
15	3	13	6			No. 15	62	28	22	10
16	4	15	7			No. 16	66	30	28	13
17	4	18	8		No. 24. Swarm taking the place of parent hive (No. 17) swapped with relocated No. 20	No. 24. Art. swarm	88	40	44*	20*
18	2	24	11	Little brood, but compact		No. 18	22	10	-18	-8
19	4	15	7			No. 19	88	40	51	23
20	4	13	6			No. 17. Parent hive	40	18	0	0
						No. 3. Relocated	35	16	0	0
						No. 7. Relocated	66	30	29	13
						No. 11. Relocated	68	31	22	10
						No. 20. Relocated	66	30	22	10
						Total harvest			516	235

* At harvest time, 18 lb (8 kg) of honey was taken from hive No. 24 and given to hive No. 18. The net harvest from hive No. 24 was therefore 44-18=26 lb (20-8=12 kg).

166. Tracking your apiary with a table

A table like the one above isn't only good for clearly summarizing the state of your apiary—it also gives precise information on procedures already completed, and those still required.

Clearly, a beekeeper should also track the progress of flowering honey plants throughout the season, along with that of each colony, in terms of its outside activity level, the number of fanning bees, etc.

Still, there are situations when a hive inspection may be required, even though the table may not have predicted it. For example, if the honey flow proves extraordinarily strong, then the most active hives may become filled with honey. In this case, you should make sure this is true by inspecting several strong hives: if they are almost full of honey, remove some honey frames and replace them with empty comb.

Simple exterior monitoring of your hives may also reveal when a hive has become queenless—if this happens, proceed as described in § 138.

In general, such overall monitoring of your apiary throughout the season won't take you more than a few minutes each morning, and will allow you to deal promptly with any unforeseen circumstances that may arise.

The honey harvest figures shown in the table represent results in an average year in a region with fairly good nectar resources. But keep in mind that these numbers may vary considerably depending on the particular year and region.

167. Fall inspection. Harvesting honey and preparing for winter

Each of these procedures should be performed just as in previous years. Only one difference may arise: if this year was fairly good, then your harvest may be much higher than the year before.

Note on extracting heather honey. If much of the honey in your region is derived from heather, you may be unable to extract the honey using an extractor, since heather honey's consistency is too thick to allow it to flow out of the cells.

1. You can leave all or some of these heather-honey frames in the hive, to serve as winter stockpiles.
2. If there is too much honey to simply leave it in the hive, and you don't want to use it to make mead, then you'll have to extract it using a honey press—in which case the comb will be destroyed.
3. If there is too much heather honey, it's best to do as follows— this will allow you to save the comb and turn the heather honey (which is always of mediocre quality) into good mead.

Uncap the combs and soak them in lukewarm water, moving them around from time to time, until almost no heather honey remains in the cells. Then, after running these near-empty frames through the extractor, put them back into the hive for the bees to clean. The honey water that results from this procedure can now be used to make mead.

168. Capped honey reserves

If you've had a strong honey season, you should resist the urge to extract all of the honey from the frames you've harvested. It's highly important to have a *frame reserve*, with frames containing capped honey, stockpiled in your honey room. Indeed, this stockpile of honey-filled frames will simplify procedures in all kinds of circumstances, and help secure the future development of your apiary.

We can't overstate the importance of this point, since a beginning beekeeper will find it hard not to give in to the temptation to harvest too much honey.

Ideally, your frame reserve should always include at least 11 lb (5 kg) per hive at the end of each foraging season.

If you can muster the foresight—or, if you like, the willpower—to establish a honey reserve along these lines, you'll avoid all of the inconvenience and worry of feeding in the spring and fall, prevent robbing—and, during a bad year, save your apiary, even if it means using up your entire reserve. So setting up a reserve has the two-fold advantage of simplification and security.

169. Inspecting comb

When, after harvesting honey, you place the freshly extracted frames back into the hives, it's a good idea to look over the comb.

Put the following aside, to be melted down later:
1. Any comb that is too irregular in shape.
2. Comb that has too many drone cells.
3. Comb that has too many pollen cells, or too many cells that are too dark.

Alternatively, instead of melting down combs with drone cells, you can simply cut them out and replace them with bits of worker-cell comb taken from another frame.

If you use the method we've suggested, there's no need to worry about leaving a certain number of drone cells in these combs, since the bees will build plenty of them on the primed frames given to them in spring (§ 161).

170. What to do during winter

During the winter season, you'll have the time to adjust the frames as just described, or to make wax (§ 277) with any comb that can't be saved. It's also a good idea to make use of this downtime by priming a number of frames using worker comb, as described above (§ 100). For this purpose, you can use the best bits of any worker comb you're planning to melt down.

Winter is also a good time to scrape off the frames to remove any excess propolis that may be found there, to build some new moveable-frame hives, and to look over all of your beekeeping equipment.

Summary

What to do in year three

Since by the beginning of year three you should have installed all of your colonies in moveable-frame hives, and since the procedures in year three are the same as in previous years, the summary of this chapter can also serve as a comprehensive summary of all of the procedures required for managing your apiary.

CHAPTER 12. YEAR THREE

SUMMARY OF BEEKEEPING PROCEDURES (SIMPLIFIED METHOD)

I. END OF WINTERING AND THE SPRING INSPECTION

1. End of wintering. For each hive, remove all of the wedges, and, after replacing the two mouse guards on the two entrances with metal gates, completely close the entrance on the side opposite the bee cluster.

2. When to complete your spring inspection. You should only carry out this inspection during nice weather, around a week after the bees have begun working actively.

Visit each hive, one by one, to determine their condition and to prepare each colony for the foraging season to come.

3. Organizing each colony. Frames containing brood should be placed in front of the open entrance; to the left and right of these frames, place those containing empty worker cells. Next, uncap any honey cells found near the top of the brood frames. In total, these frames should number between ten and twelve. Then, fill the rest of the hive with honey frames, alternating them with empty primed frames.

4. Queenless and disorganized hives. Any such hives you find will need to be dissolved.

II. MAINTAINING OR INCREASING THE NUMBER OF HIVES

1. Artificial swarming. Place all of the frames from a strong hive into an empty one, leaving the bees behind in the original hive, now filled with primed frames. Then, place the new hive, now filled with brood comb but containing no bees, in the former location of a second strong hive, which should be moved elsewhere.

2. Hives bought some distance away. You don't have to rely solely on artificial swarms to increase your number of hives—from time to time, you can also buy new hives, from some distance away.

III. MONITORING THE APIARY DURING FORAGING SEASON

1. Tracking your apiary using a table. Create a table outlining, at a glance, the condition of your colonies, providing information on the ways in which particular hives should be monitored in order to avoid unforeseen circumstances.

2. General external monitoring. Between the spring inspection and the honey harvest, you should conduct frequent external inspections of your apiary's overall condition—a process that should not require much of your time.

IV. Honey Harvest and Wintering

1. Harvesting honey. Remove the honey frames, for the most part capped, and redistribute some of them as necessary among your various hives to ensure that each has honey stockpiles of approximately 35 lb (16 kg).*

A certain number of the honey frames removed from the hives should be set aside for your *capped honey reserve*; the others can be extracted.

2. Wintering. Put the extracted frames back into the hives, and, before the first frost, replace the hive entrance gates with mouse guards. Finally, insert some suitable wedges in between each hive and its bottom board, and between each bottom board and its stand.

---

* Larger reserves may be needed in northern climates with longer winters. See comment to § 125. *Ed.*

PART THREE
OTHER HIVE SYSTEMS

Figure 153. An apiary in the Cévennes Mountains.

Chapter 13

Vertical Hive Equipment

171. The vertical movable-frame hive

Up to this point, we've assumed that you are using the French ("Layens") moveable-frame hive, which is the simplest in form and the easiest to use. Our preferred model is this *horizontal* hive, so called because all of its frames are arranged in a single row. This means that the colony is expanded horizontally, by extending the row of frames.

But there's also another type of hive that is called the *vertical hive*, since this system allows you to stack multiple tiers of frames, one atop the other, thus expanding the colony vertically, upward. Such vertical hives are also referred to as *hives with supers* (figs. 154 and 157).

This hive system is based on the principle that bees in a fixed-comb hive that is taller than it is wide prefer to store their honey in the upper part first. As we've seen with the cap hive, some beekeepers had the idea of making this upper section a separate component, containing nothing but honey, while the lower part contained the brood and any remaining honey. So the cap is a kind of super, which makes it easier to harvest the surplus honey. Applying this principle to the moveable-frame hive gave birth to the vertical hive.

A vertical hive or hive with supers consists of several stackable boxes; the lowermost box, called the *hive body* (B, figs. 154 and 157), contains the frames (*bf*, fig. 154) that are dedicated exclusively to raising brood and storing the winter honey reserves.

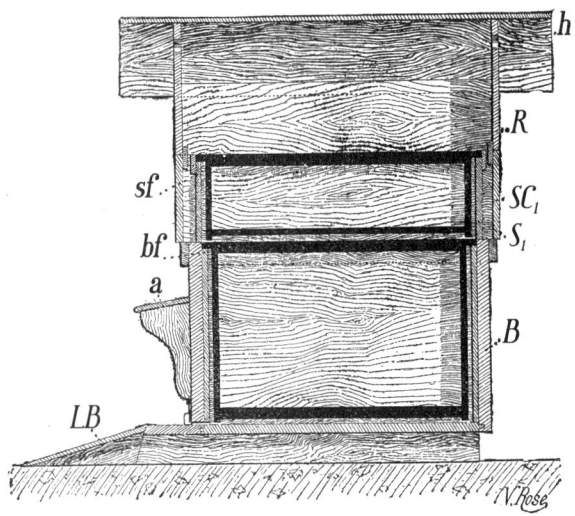

Figure 154. A vertical hive with a single super (cross-section, front to back): B, hive body; *bf*, a hive body frame; SC_1, super cover; R, roof; lb, the landing board; *a*, entrance awning; *h*, one of two holes with wire netting.

These frames are wider than they are tall or, as beekeepers say, "*longer than they are deep*" (the horizontal dimension, or *length*, is greater than the vertical dimension, or *depth*)—the opposite of those in the horizontal hives we looked at previously. This shallower frame design is meant to encourage the bees to make their way up into the upper compartments of the hive.

The remaining boxes are called *supers* (S, fig. 154), and fit snugly atop the hive body, or atop other supers. The supers contain the frames (*sf*, fig. 154) that exclusively contain the surplus honey that will eventually be harvested.

These super frames are approximately half the depth of those in the hive body; the very shallow frames in the supers allow the bees to store their honey more quickly.

This design—more complicated than that of the horizontal hive—is meant to keep the honey that will be harvested completely separate from the rest of the colony, which will spend the winter in the lower box after the supers have all been removed.

172. A description of the vertical hive

Just as there are many models of horizontal hive, there are also many varieties of vertical hive.

Among the latter, we'll be describing the following:*

1. *Hive body.* Aside from its dimensions, the *hive body* (B, figs. 154 and 157), is constructed just like a 12-frame horizontal hive, but with only one entrance, positioned in the middle.

Each frame is longer than it is deep (fig. 156); a frame has an interior depth of 10 5/8" (27 cm), and an interior length of 16 1/2" (42 cm).

The hive body is set atop a bottom board that fits under the hive, and whose outer section is sloped, forming a kind of deck (*LB*, figs. 154 and 157).

When there is no super placed atop the hive body, the top of its frames is often covered with an *oilcloth* that is removed when supers are added. When supers are present, this oilcloth is placed atop the uppermost super (*c*, fig. 157).

2. *Supers.* A *super* is a box with neither a bottom nor top, containing twelve frames that are approximately half as deep as those in the body (fig. 155): these frames have an interior depth of 5 5/16" (13.5 cm) and an interior length of 16 1/2" (42 cm).

The bottom of a super can fit onto the hive body, or be placed atop another super—and that same super's top can have another super set atop it. The lowermost super, S_1, is protected by an outer casing, SC_1 (figs. 154 and 157).

---

* This hive, referred to as a *Langstroth hive* in honor of its inventor, has seen several modifications of its dimensions; commercially, it's known in France as a *Dadant hive*, the latter being a less fitting name than the former.

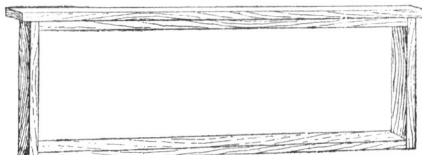

Figure 155. A super frame from a vertical hive.

Figure 156. A brood frame from a vertical hive.

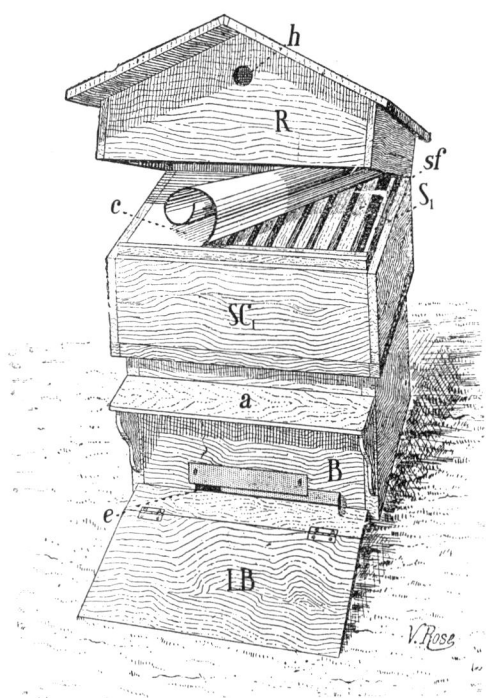

Figure 157. A vertical hive with a super: *B*, hive body; SC_1, cover of the first super; S_1 super; *sf*, frames of this super; *c*, oilcloth covering the frames; *R*, roof; *h*, ventilation hole with wire netting; *a*, entrance awning; *e*, entrance; lb, landing board.

In a region with sufficient honey flow, each hive of this kind should have at least two or three supers.*

3. *Roof.* The *roof* (R, figs. 154 and 157), is a top that fits atop either the hive body or any super. The roof has two screened ventilation holes, on opposite ends (*h*, figs. 154 and 157).

So the entire hive may consist of nothing but the hive body with the roof on top: this is what it looks like in winter, for example. Or, the hive body may have one or several supers stacked atop it, covered by the roof: this is what it looks like during heavy honey flow.

173. Notes on the vertical hive

As we can see, the vertical hive's capacity is variable. In a horizontal hive, the bees themselves are left to determine how much space they occupy. Here, the beekeeper intervenes, increasing or decreasing the hive's capacity by adding or subtracting supers.

The main advantage of this capability is that the honey harvest can be done all at once, as with the fixed-comb cap hive.

However, the vertical hive is more difficult to manage, since it requires beekeepers to add or subtract supers themselves at the appropriate times—and you may not always get the timing right. With a horizontal hive, meanwhile, the bees are allowed to expand at their own discretion, and they have a knack for taking up the right amount of space based on external circumstances.

Another overall difficulty is due to the fact that the super frames are only half as deep as those in the hive body. So you won't be able to exchange these two types of frame with each other as you perform your various procedures.

So, as you can see already, it takes an experienced and skilled beekeeper to manage bees using vertical hives.

As for honey yield, we can say that, all else being equal, a vertical hive will yield roughly as much as a horizontal one.**

---

* This is an important point to remember when buying vertical hives, since manufacturers' catalogs list prices for vertical hives with a single super; if you need to buy a hive with at least two supers, the price will rise by approximately 5 francs, and is always higher than that of a comparable horizontal hive with the same capacity.

174. The advantages of a vertical hive for producing section honey

This kind of vertical hive is more conducive than any other for the production of *honey sections*—that is, honey to be sold in comb inside small wooden frames (fig. 158).

The honey contained in these cells, newly built and capped—resulting in a comb that is completely joined along its edges to the interior of a small wooden frame—comes in an appealing form that's perfect for desserts.

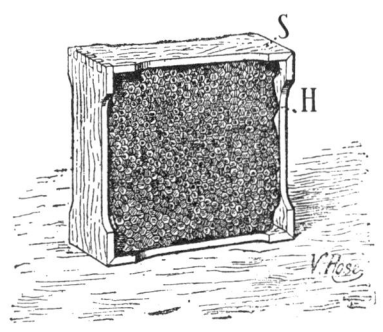

Figure 158. A finished section: *S*, section; *H*, honey.

But a section like this, containing good honey, is always a bit of a luxury item. It has been proven time and again that bees are put to a lot of trouble to build comb in these sections, since when the colony is split up into multiple working groups, ventilation and temperature regulation are hindered by the dividing walls between all of these small section boxes.

In fact, if we compare the honey harvest in hives with sections and those containing simple frames—both from the same region and at the same season—we find that the hives with sections only yield half, or even a fourth, as much honey by weight as a frame hive. This means it's more profitable to sell extracted honey at 1 franc per kg than section honey at 2 francs per kg.

It's also worth noting that producing nice-looking sections that are attractive to buyers can be very time-consuming: you'll need to monitor them, change their position, and pay attention to all kinds of minute details (§§ 191–193).

Finally, transporting sections presents serious difficulties: if they're

---

** However, Beuve, who compared results for 12 vertical hives and 12 horizontal hives over a period of 10 consecutive years, found that horizontal hives were slightly more productive, but the difference was negligible (see "Comparaison du rendement des ruches horizontales et des ruches verticales," by Beuve, President of the Société d'Apiculture de l'Aube (in *Apiculteur*, 1895, p. 375).

CHAPTER 13. VERTICAL HIVE EQUIPMENT

not wrapped in a particular way, they run the risk of being damaged by any sudden jolts they may receive.

For all of these reasons, it's generally not in your best interest as a beekeeper to produce section honey, unless you make this your specialty, and unless you're certain that they'll fetch a sufficient price to make it worth your while.

175. Equipment needed for producing section honey

The sections normally used are small frames made with strips of wood that are thinner and wider than those used in ordinary frames—thinner, to lend them a more elegant appearance and a lower weight; and wider, since, to ensure that the comb is built uniformly, the distance between the edge of the section and the surface of the comb, on all sides, should equal the thickness of a single bee. Figure 159 shows an unassembled section. An assembled section—built out by the bees—is marked S (fig. 158).

Ordinary sections are designed so that the final result weighs approximately 1 lb (500 g), wax and wood included.

So to achieve uniformity in comb construction, the bees must be

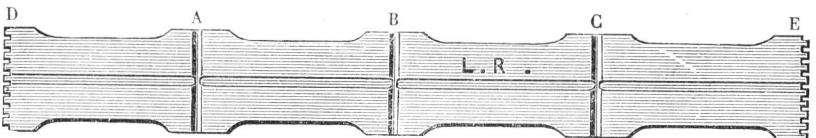

Figure 159. An unassembled section. To assemble it, moisten the three grooves and fold them at the points A, B, C; then, using a hammer, join points D and E by interlocking the notches.

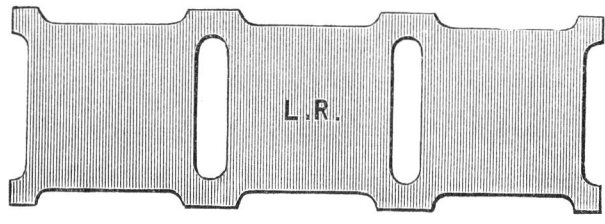

Figure 160. A triple separator made of sheet metal, used for three sections at a time.

deterred from excessively elongating their cells. This led to the invention of so-called *separators*. These are strips, made generally of tin (figs. 160 and 161), that are inserted between the sections.

These separators and the sections' wooden frames are notched at the top and bottom to allow bees to pass through, moving freely between sections. The sections are usually placed next to each other in boxes (fig. 162), held in place by wooden screws.

The section boxes fit squarely atop the frames of the hive body when the oilcloth has been removed—and the boxes themselves are covered by an oilcloth (*c*, fig. 162).

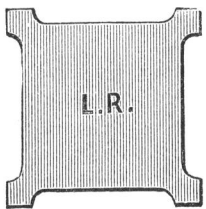

Figure 161. A single separator.

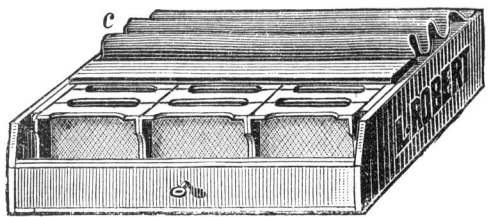

Figure 162. A section box. You can see the gaps through which the bees can move between sections. Triple separators have been placed vertically in between each row of three sections, but the separators are not visible here; *c*, oilcloth.

Summary

Vertical hives

The *vertical hive* or *hive with supers* is, in essence, a hive body containing a dozen frames (longer than deep) atop which a roof can be placed.

In between the hive body and the roof, you can insert one or several supers, each containing a dozen frames that are as long as those in the hive body, but only half as deep.

Section honey

One advantage of the vertical hive is that it makes it easier to harvest honey, and is good for producing *section honey*—that is, honey that is still in the combs, built in small wooden frames. But producing such honey is more complicated and requires a beekeeper with a lot of experience and skill.

Section honey is only profitable if you can sell it at a much higher price than simple extracted honey. The sections are arranged in boxes and placed atop the frames of the hive body.

Chapter 14

Managing Bees In Vertical Hives

176. General observations

Almost everything we discussed in part two of this book—from setting up your apiary, springtime procedures, swarming and installing swarms in hives, to fall inspection and wintering—holds for managing vertical hives as well. However, in certain respects you will need to proceed somewhat differently.

Let's assume that you have several populated vertical hives and track these colonies for an entire season, from the end of one wintering period to the start of the next, and point out all of the procedures that differ from those described for horizontal hives.

177. The end of wintering and the spring inspection

During wintering, vertical hives are exactly like twelve-frame horizontal hives, consisting of nothing more than a hive body covered by a roof.

Once you've removed the hives' special winter set-up, inspect each one at the proper time and take note of their condition.

For each hive, take care to leave the brood in the middle frames, while moving the frames with the most honey to the edges, both right and left. Then, fill the rest of the hive with comb made of worker cells, either empty or with little honey—this arrangement will allow the queen to lay her eggs unimpeded. During this inspection, remove honey frames from the colonies that have the most, and give them to the ones that have too few.

178. Preparing supers

You should prepare supers with frames in advance—soon you'll need to place them atop your hives.

Once you've had vertical hives for several years, you'll have enough comb-filled frames to equip your supers.

If you're preparing supers for new hives, the frames should all be primed with foundation, except for one frame near the middle of the super—this frame should be entirely filled with comb in order to encourage the bees to climb up into the supers.

Frames primed simply with bits of comb won't be good enough: since the supers will only be set in place a few days before the peak foraging season, the bees will need enough available comb to take full advantage of the honey flow.

However, if you're being frugal, you can use the following arrangement, whose results are almost as good. In the super frames, put pieces of foundation sheet, cut into triangles like that marked *A*, *B*, *C* in figure 163. The triangular piece of sheet is attached on the sides *AB* and *BC*, with the remaining side *AC* forming a diagonal.

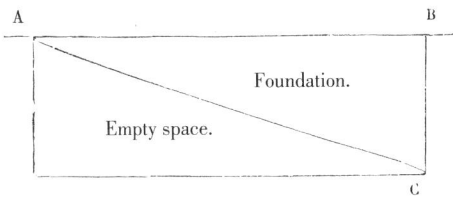

Figure 163. The most economical way to add foundation to supers.

179. When to add the first super

It can be difficult to pinpoint the right time to add the first super—doing so requires a deep familiarity with the plants and climate of your region, allowing you to predict the approach of the main honey flow based on atmospheric conditions.

The best time to add the first super is several days before the main honey flow begins.

But before you add the super, there's one other condition that should be met: almost all of the frames in the hive body should be full of bees. Any hives that are too weak should be united, in pairs—otherwise they won't be strong enough to take advantage of the honey flow.

If you don't unite your weak hives, you'll need to wait for them to grow stronger before adding the first super, but when you finally do so it may well be too late—that is, after the main honey flow is over.

180. Adding the first super

To add the first super, remove the roof, lift the oilcloth on one side, and continue to smoke throughout the hive as you gradually lift the rest of the cloth.

Fit the super, filled with frames, atop the hive body, put the oilcloth back atop the frames of the super (*c*, fig. 164), and replace the roof, *R*, on top.

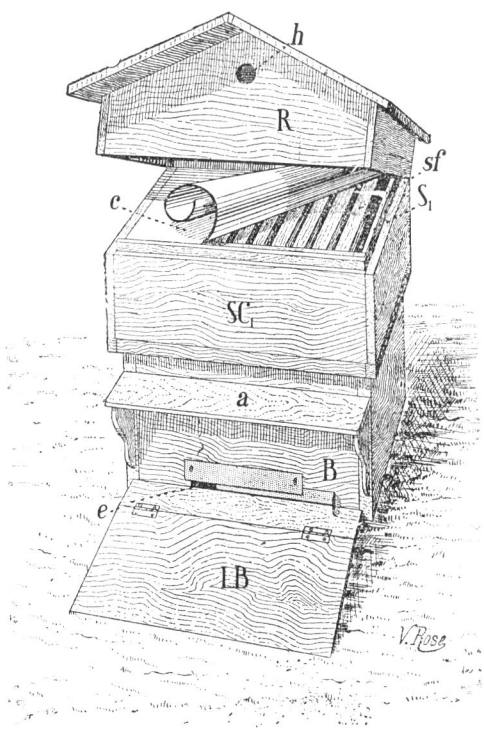

Figure 164. A vertical hive with a first super added, S_1; the oilcloth, *c*, that previously covered the top of the hive body now covers the first super. (The remaining components are marked as in fig. 157.)

181. Problems that can result from adding the first super too early

If you add your supers too early, then any sudden dip in the temperatures, combined with the sudden expansion in the size of the hive, may cause the bees to abandon part of the brood when their cluster contracts.

Any brood that is no longer covered by the bees could perish, and even lead to an outbreak of foulbrood disease, which poses a tremendous threat to any apiary.

182. Problems that can result from adding the first super too late

If you add your supers too late, then:
1. You'll fail to take advantage of the entire honey flow period.
2. If the hive's population exceeds the capacity of the hive, they'll grow likely to swarm—and we've already seen the problems that can result from natural swarming.
3. If the queen runs out of room to lay eggs due to all the honey being stored in the hive body, she may climb into the super to lay there instead—and if there's brood in the super, it won't make sense to harvest the supers until the brood has completely emerged.

183. Monitoring your supers

Since the intensity and duration of the honey flow may vary from year to year, and the supers may be filled at various rates depending on these circumstances, you'll need to monitor the supers on all of your hives in order to determine the best time to add your second round of supers.

Some beekeepers simplify such inspections by using supers with glass walls on one side, with the glass covered by a shutter that can be fastened shut with a hook.

184. Adding the second super

If the honey flow is sufficient, the first supers will begin to be filled with honey—but you don't have to wait until they're completely full,

CHAPTER 14. MANAGING VERTICAL HIVES

since, as we know, bees require a rather large amount of comb for driving off the excess water from the honey before capping it.

The best time for adding the second super to a hive is *when the first super is two-thirds full*. But this second super (S_2, fig. 165) isn't placed on top of the first, but rather inserted in between the hive body and the first super—that is, where the first super stood previously. The first super, S_1, is placed atop the second super, S_2, and the oilcloth, c (fig. 165), is left where it is.

This method has the advantage of leaving the empty frames close to the hive body, encouraging the bees to fill them more actively.

Before adding the second super, it's a good idea to inspect the first to see whether or not it contains any brood. If it does, you'll need to place the second super on top of the first—if you place it underneath, then the queen will lay in the second super even more than in the first in order to bridge the two parts of the brood nest that have been separated.

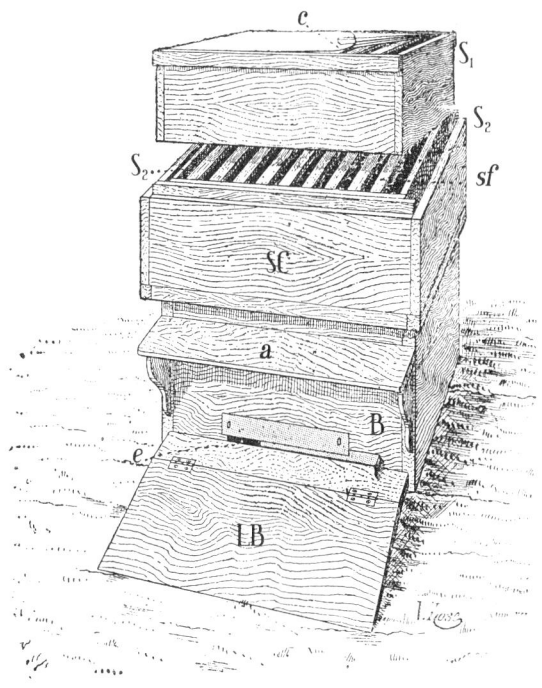

Figure 165. Adding the second super S_2, in the position previously occupied by the first super. The first super, S_1, is placed on top, and the oilcloth, c, remains on top of it. (The remaining components are marked as in fig. 164.)

185. More supers

In regions with particularly intense honey flow, when the bees continue to forage, you'll need to insert a third super beneath the second. In unusually good years, even more supers may be needed.

186. Adding supers for the autumn honey flow

In regions where the primary honey flow comes at the end of the season, one should, of course, add the supers at the end of summer and a few days before the expected main honey flow.

In regions with both spring and autumn honey flow, you should add supers in the spring, then harvest them, before putting them back in time for the autumn honey flow, after which you can harvest them once again. These numerous procedures do have the advantage of allowing you to better separate various kinds of honey produced during the various periods.

187. Inspecting the hives when supers are in place

If, in order to perform some procedure with a hive or check on its condition, you have to inspect it while supers are in place, then do so as follows: remove the supers, place them on the ground, and, before inspecting the hive body, smoke intensely.

If you're forced to perform this procedure at a time when there is no nectar in the flowers, you'll need to take steps to prevent robbing.

188. Harvesting the supers

Don't be in any great hurry to harvest your supers, since the longer the honey remains there, the more cells will be capped, and their contents thus better preserved. You can harvest a super as follows.

Remove the roof and, smoking heavily, drive some of the bees down toward the hive body,* then quickly inspect the super to make sure it doesn't contain any brood—if it does, you'll need to leave the super in place until all of the brood emerges.

If there isn't any brood—and there usually isn't—detach the super along its bottom using a chisel; smoke once again, then remove the super and place it for the time being on a stool. Then, put the oilcloth back on the lower super or the hive body, and promptly replace the roof.

Carry the supers one by one into your honey room, and prop each up on a wedge and cover the top with some cloth.

Any bees that remained inside the supers will sense that they've been separated from their hive and will gradually leave the supers through the bottom and, seeking to rejoin their colonies, fly toward the honey room windows, which you should open from time to time to let the bees out.**

The honey frames are then removed from the supers and extracted.

Next, put the now empty frames back into the supers, and, that evening, put the supers back atop the hives to allow the bees to clean the frames. To avoid robbing, narrow the entrances of all of your hives for several days.

189. Inspecting after harvesting

It goes without saying that you should inspect all of the hives after the harvest; during this inspection, check to see that each hive body contains approximately 35 lb (16 kg) of honey.

Since in vertical hives most of the honey is found in the supers, their hive body often does not contain enough honey.

So it would be only natural, come fall, to take some honey frames from the supers and move them into the hive body to fill out the

---

* To encourage the bees to climb down into the hive body, you can use a *carbolated cloth*. Pour some carbolic acid—enough to fill a small wine glass—into a bucket of water. Soak the cloth in this solution, and, stretching it out, place it on the super. The smell of the carbolic acid will prompt a good number of the bees to descend into the hive body. (The products used today for this purpose include Fischer's Bee-Quick or essential-oil-based Honey-B-Gone on a fume board. *Ed.*)

** Some have recommended a special device called a *bee escape* (§ 226) for use in harvesting supers; but, all things considered, the method we've described is quicker.

winter honey reserves—but since the super frames are of a different size than those in the hive body, this is impossible.*

Meanwhile, since there is rarely a sufficient surplus in the hive bodies of vertical hives to yield a sizeable reserve of these larger frames, it's simply not realistic to try to establish a reserve.

Experienced beekeepers can often avoid fall feeding, with all of its potential problems, if they manage to remove the supers at the right moment—namely, shortly before the end of the main honey flow. This way, the bees will be storing honey in the hive body for the rest of the foraging season.

190. Preparations for wintering

As we have said, vertical hives should not include supers during winter.

You might think that if there aren't enough honey stockpiles in the hive body, you could simply leave a super full of honey in place to ensure a sufficient reserve. However, if the winter includes prolonged cold periods, then the bees will be unable to move about the hive, in which case the honey-filled super would be of no use to them.

In fact, during winter, the cluster of bees in the hive body would be unable to climb into the super because of the gap that inevitably separates the combs of these two sections of the hive.

As for preparing the hive body—with the roof on top—for wintering, you can proceed just as you would with a horizontal hive.

191. Sections. Comb honey

You may want to produce several units of section honey (see § 174) for your own consumption, to grace your table with honey in this elegant form—for which purpose just a few comb sections will suffice. In this case, you can also produce some sections using horizontal hives (§ 194).

---

* There are vertical hives that use identical frames in the supers and in the hive body, but, as we saw earlier, such supers are more difficult to fill. Another option, proposed by Brother Jules, is to keep horizontal and vertical hives in your apiary simultaneously, with the same frame dimensions.

If, on the other hand, you'd like to produce section honey in greater quantities in order to sell it, and if you're certain they'll turn a profit, then you can use vertical hives with shallow frames like those described above.* In order to do so, however, you'd need to make this your specialty, since producing sections of comb honey involves—as we've seen—complicated equipment and painstaking and extremely time-consuming work.

On top of all this, a beekeeper who specializes in producing sections isn't always guaranteed to succeed, since it's no easy task to put your bees to work in these cramped compartments, forcing them to build complete and impeccable combs—propolis-free, with all their cells capped, and attached uniformly to the section walls on all four sides.

And it's hard to sell sections that don't combine all of these qualities.

192. How to encourage the bees to fill out the sections with comb

After moistening the section at the grooves A, B, and C (fig. 159) so it won't break when bent, fold it around a block of wood, then staple the ends D and E together.

Into each section, insert foundation—thin and white, made especially for this purpose. You can get by with simply priming each section with a strip of regular foundation, but, to insure that the comb is built uniformly, it's preferable to prime each section in the center with a piece of foundation of the same dimensions as the interior of the frames. When doing so, be sure not to attach the foundation sheet using wire, since the comb of each section is to be consumed as is, in its entirety.

To attach the foundation sheet inside the sections, you can, for example, use a block of wood (b, fig. 167) that has the desired dimensions and is just under half as thick as the section itself; place the section around the block and set the foundation sheet on top of the

---

* The shallow-frame vertical hive was invented precisely in order to harvest sections, which were once referred to as *surplus boxes*. In the United States and England, honey is usually sold in this form.

block (*f*, figs. 166 and 167), then attach it to the section uniformly on all sides by pouring molten wax on it. To pour the wax, which you should melt using a water bath (a "bain-marie," as discussed in § 221), tilt both the section and the wooden block, supporting the latter with your left hand, and use your right to pour a thin trickle of wax into the angle formed by the foundation sheet and one side of the section—then do the same for the three remaining sides.

Now that they've been primed with foundation, place the sections into the boxes with the dividers and use a wooden screw to squeeze them tightly together in order to prevent, as much as possible, any propolis buildup around the edges.

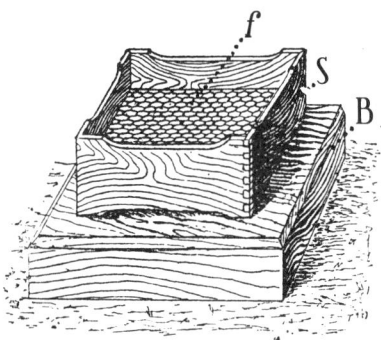

Figure 166. Adding foundation to a section: *f*, foundation; *S*, section; *B*, block of wood, supporting a smaller block (*b*, fig. 167) that reaches halfway up the section, *S*.

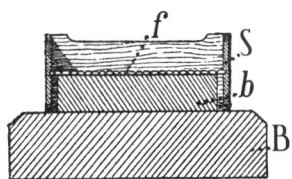

Figure 167. Section and wooden blocks, cross-section from top to bottom. You can see the block, *b*, that supports the foundation, *g*, being added to the section.

Place the boxes, now full of sections (fig. 162), atop the hive body, and, after placing the cloth on top, put back the roof. The section boxes should be placed just like supers—that is, a few days before the main honey flow begins.

Once the boxes are in place, monitor them as follows.

1. Since the bees start by filling the sections in the middle before moving on to those on the sides, you'll need to remove those in the middle as soon as they're completely finished and replace them with the unfinished sections to each side, replacing the latter, in turn, with the sections that are still completely empty.

2. It's important to remove the sections as soon as they're finished, since the bees will stain the capped comb if they're allowed to remain on them for too long.
3. Scrape clean any sections you remove that have too much propolis. This is a delicate procedure, requiring a light touch.

193. Problems to avoid when producing section honey

An initial difficulty is that the bees may sometimes hesitate to climb up into the sections. You can encourage them by replacing several sections in the middle with some that have been primed with completed comb instead of foundation sheet.

Beekeepers have attempted to speed up the bees' passage into the sections by reducing the number of frames in the hive body—for example, by leaving only 7 or 8 frames in the middle, separated from the rest of the hive body to both right and left with division boards (see § 227). But there are two drawbacks to this technique. First, the bees are often reluctant to build in the sections located to the right and left above the empty parts of the hive body. Second, and more importantly: even if your sections come out nicely, you still risk leaving too little space, after narrowing the hive body, for proper brood development and winter reserves—which could endanger the colony in the future.

Another problem is that hives with sections tend to be more likely to swarm, which, as we know, causes all kinds of trouble. There is no practical way of avoiding this problem.

Finally, if the section box isn't placed at the right time, the queen may climb up out of the hive body and lay eggs in the sections, ruining them completely.

194. Section comb honey with horizontal hives

You can also produce comb sections using horizontal hives, as follows:

1. After preparing the sections as described above, place the box containing them (fig. 162) beneath the roof of the horizontal hive, positioning it above the brood frames. First, be sure to remove the

slats placed in between the frames or the boards covering them, in order to allow the bees to travel up to the comb honey sections.

The horizontal hive we've described allows for a section box to fit under the roof. Whenever you're adding sections to horizontal hives, be sure to choose the strongest ones.

2. You can also arrange the sections vertically, in a frame prepared especially for this purpose (fig. 168)—but one problem with such frames is that the bees tend to work more near the top. Place the section frame alongside the final brood frame.

Figure 168. Frame readied for vertical placement of sections in a horizontal hive.

Summary

Managing vertical hives

When managing colonies in vertical hives, the required procedures involving the hive body are more or less the same as those required for horizontal hives.

However, there are a few additional procedures unique to the vertical hive: adding the first super to each strong hive shortly before the main honey flow; inserting the second super in between the first one and the hive body, when—and only when—the first is at least two-thirds full of honey; monitoring the supers and inspecting them during the honey harvest to make sure they contain no brood; harvesting the supers, moving them to an enclosed room after driving most of the bees off of them, then removing the frames for extraction.

During your fall inspection of the hives, take note of which colonies are running low on honey for the winter, and will need to be fed or united with other hives.

Section honey

When you wish to produce some section comb honey for your personal consumption, you can use horizontal hives by adding section boxes beneath their roof. However, if you're planning to produce sections for commercial sale, it's better to use vertical hives. Producing sections presents numerous difficulties and is labor- and time-intensive. To do it successfully, you should have enough free time at your disposal, and gain the experience necessary for this specialized kind of beekeeping.

Vertical hive advantages and limitations
In sum, vertical hives have several advantages: first, harvesting honey takes less time; second, they allow you to keep the various honeys produced during various seasons separate; and, third, they are the only hives capable of producing comb honey on a commercial scale. However, these hives are more difficult to manage than horizontal ones. On top of that, vertical hives make it more difficult to avoid swarming and fall feeding.*

---

* Surplus honey frames in the supers are not the same size as brood chamber frames, which makes it difficult to add frames of honey to the brood chamber. In addition, the fall honey flow after honey supers are removed may not be adequate to build up winter reserves. *Ed.*

Chapter 15

Keeping Bees in Traditional Fixed-Comb Hives

195. General remarks

In chapter 5, we assumed that a beginning beekeeper would learn to handle bees and carry out basic beekeeping procedures using fixed-comb hives. There's no need to revisit such questions here: we've already discussed how to buy fixed-comb hives, how to assess their condition when buying them, how bees winter in such hives, how to inspect them in spring and fall, how to collect any swarms they produce, how to prevent robbing, and how to feed fixed-comb colonies.

But we also assumed that the beginner would only own and use such fixed-comb hives in order to turn them sooner or later into moveable-frame hives.

Sometimes a beekeeper who has started out using fixed-comb hives may hesitate to turn them into moveable-frame hives due to the expenses associated with this process, and may therefore continue using fixed-comb hives exclusively. Although, given the ease and other advantages of keeping bees in moveable-frame hives, such a beekeeper may decide—during a particularly good harvest, for example—to use some of the money generated by the fixed-comb hives to purchase two or three moveable-frame hives.

However, you may also choose to continue using fixed-comb hives only—if, for example, it's more profitable to sell your occupied hives than your honey. In that case, you can pursue *bee breeding*, which can be advantageous if you're not in a particularly nectar-rich region.

CHAPTER 15. TRADITIONAL FIXED-COMB HIVES 219

So you'll start out using fixed-comb hives in any event, and continue using them if you can't yet afford the moveable-frame hive equipment. Indeed, you may continue using them regardless, if you decide to specialize in breeding—that is, selling populated hives.

Now we'll fill in any details missing from our earlier discussions of fixed-comb hives concerning how to keep bees using traditional methods.

196. The end of wintering and springtime procedures

Let's assume, for example, that you prepared fifteen fixed-comb hives for wintering the previous fall, and take a look at the procedures you'll need to carry out between the end of one winter and the start of the next. The wintered hives should be handled as described in § 78 and onward, and each should be inspected carefully (fig. 169).

Let's say that the inspection yields the following results:
1. One dead hive.
2. One disorganized hive.
3. One strong hive that is running low on honey.

Figure 169. A beekeeper smoking a fixed-comb hive before inspection.

4. Two weak hives that have wintered successfully.
5. Ten hives in excellent condition.

In this situation, you'll need to:

1. Dissolve the dead hive, treat it with sulfur as described in § 86, and save the comb, if it's not too old—after cutting out brood and drone cells—to house a natural swarm later. If the comb is too old, sulfurize the hive, remove the sections of comb containing brood, and wait for the harvest season to harvest the hive's honey and comb while doing the same for your remaining hives.
2. Put the bees saved from the disorganized hive in other hives as described above (§ 85), then handle the empty hive just as you did the dead one.
3. Feed the strong hive running low on honey as described above (§ 87).
4. Wait until the honey flow begins to swap out the two successfully wintered weak hives with strong ones.
5. The strong hives in excellent condition should be monitored throughout the season, since one of them might grow weak by the time the honey flow begins—in which case it should be swapped with a strong one, as we just mentioned.

197. Swarming season

In the simplest case—that is, when one allows primary swarms to emerge—you can collect the swarms as described beginning in § 104 and install them in fixed-comb hives.

Of course, afterswarms should always be returned to their hives of origin (§ 113).

Prime swarms that have been collected and installed in new, empty hives should be monitored, and even fed if the weather is poor.

Following swarming season, we can assume that, in an average year, our hives stand as follows:

1. A strong hive, fed in the spring, that has not swarmed.
2. Eight hives, strong in the spring, that have swarmed, and whose afterswarms were returned to them.
3. Two hives, weak in the spring, that have been swapped with strong hives that have not swarmed.

4. Two hives, strong in the spring, that were swapped with the preceding two hives, and that haven't swarmed.
5. Eight prime swarms, collected and installed in fixed-comb hives.

198. Uniting weak or late swarms

Not every swarm, however—even primary and other early swarms—should always be preserved: weak swarms should be united.*

The same is true of swarms that are strong, but come late in the season, since such swarms will not have enough time to build up sufficient winter reserves despite their large population.

We should note that when you're increasing or maintaining the size of your apiary using natural swarming, *the future of your apiary depends on uniting weak or late swarms.* Two good swarms are better than four mediocre ones.

199. How to unite swarms with swarms

Here's how to unite swarms in various situations:

1. *The two swarms did not emerge on the same day.* In this case, you should add the most recent swarm to one that was collected a few days earlier and has begun building comb.

On a smooth and even section of ground, place two sticks, about 5/8" (2 cm) thick, parallel to each other, around 8" (20 cm) apart. Near these two sticks, gently place the two swarms to be combined on the ground, each in the fixed-comb hive you collected them in.

Smoke each swarm until the pacified bees begin to hum loudly; for this procedure to succeed, it's essential to maintain this humming both before and after. Next, take the most recent swarm and shake it abruptly, causing it to fall out onto the sticks—then cover it with the other swarm. Smoke all around to drive the bees into the hive, then smoke underneath it to maintain the "humming" state. Keeping them humming will prevent the bees from fighting.

---

* Generally, a swarm should be considered fairly weak if it weighs less than 3 lb (1.5 kg).

2. *The two swarms emerged on the same day.* In this case, the weaker swarm should be added to the stronger one, in which case you can proceed more quickly than in the previous case.

After smoking both swarms until the pacified bees begin to hum loudly, shake the weaker one out into the stronger one, then cover the united hive with a bottom board, turned upside down. Next, turn the entirety right-side up, while holding the hive tightly atop the stand. Finally, smoke the hive through the bottom, propping it up as described before.

200. Artificial swarming with fixed-comb hives

If you'd like to alleviate all of the problems caused by natural swarming, and also avoid having to unite swarms, you can increase your apiary by creating artificial swarms using fixed-comb hives. This method also has the advantage of ensuring that you have swarms, even in years when they wouldn't emerge naturally—and of producing them when you want, under the best possible conditions. So artificial swarming can be recommended not only for ordinary beekeepers, but especially for those wanting to concentrate on breeding, and whose only aim is to sell populated hives.

One of the best ways to pursue artificial swarming with fixed-comb hives involves the same principle as that used with moveable-frame hives (§ 163), but it's a bit different in terms of the practicalities.

The plan is to create an artificial swarm using two strong hives that will be relocated.

Let's assume that that these two strong hives are hives A and C as marked below (fig. 171); alongside them, place the empty hive, B, in which you want to establish an artificial swarm using the two strong hives A and C.

Drive the bees from hive C into hive B (§ 146 and fig. 170), then place hive B in the location previously occupied by hive C. Next, place hive C—now mostly empty of bees—in the former location of hive A, which in turn should be moved to another area of the apiary.

Hive C will produce a new queen, thus leaving you three hives instead of two (fig. 172) at the end of the season.*

CHAPTER 15. TRADITIONAL FIXED-COMB HIVES 223

Figure 170. A beekeeper driving bees into an empty hive using the "drumming" method; the two hives are joined on one side using wire hooks.

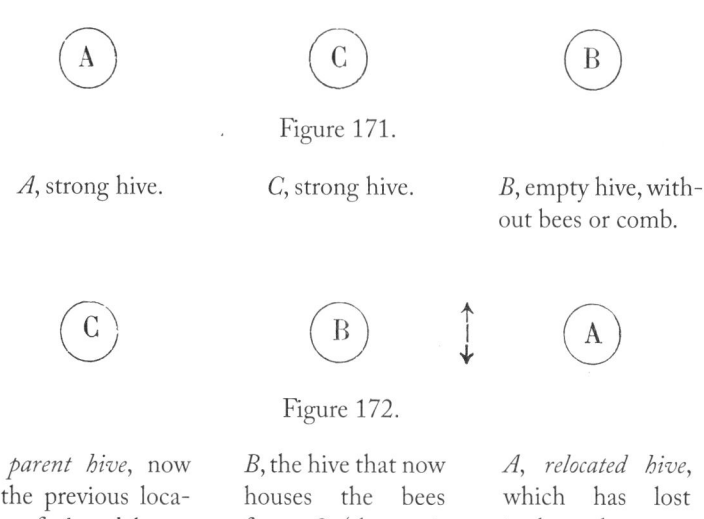

Figure 171.

A, strong hive.

C, strong hive.

B, empty hive, without bees or comb.

Figure 172.

C, parent hive, now at the previous location of A, and home to those bees from A that were away foraging. This hive will produce a new queen.

B, the hive that now houses the bees from C (the *artificial swarm*), and those bees from C that were away foraging.

A, *relocated hive*, which has lost its bees that were away foraging.

These three hives include: hive C, the *parent hive*; hive A, the *dislocated hive*; and hive B, the *artificial swarm*.

If you'd prefer to avoid natural swarming as much as possible, carry out this procedure with all the strong hives in your apiary.

Remember, such artificial swarming should take place approximately fifteen days before the main honey flow.

201. Harvesting honey from fixed-comb hives**

The best way to harvest honey from fixed-comb hives is to completely harvest a certain number of heavy hives, whose bees will be subsequently added to other hives.

Figure 173. Traditional method for driving bees.

---

* Hive C may sometimes produce an afterswarm thirteen or fourteen days after the procedure—of which the queen's song will provide warning. If this happens, collect the swarm and return it to hive C two days later (see also § 233).

** See also § 244.

If the season has seen a heavy honey flow, then the hives that have swarmed—and even those containing that year's swarms—should have more than sufficient honey reserves, and the hives that haven't swarmed should be almost full of honey.

Start by harvesting the heaviest hives; if they include hives that have swarmed or parent hives that have been used to produce artificial swarms, be sure to harvest them roughly twenty-one days after the prime swarm emerges, since by that point they will contain no more brood.

To harvest a hive, drive the bees into an empty hive, either using the method described in § 146 (fig. 170) or by attaching them with fabric (fig. 173). Move the empty hive, now containing the bees, to the location previously occupied by the hive being harvested; these bees will later be added to one of the weaker hives that wasn't harvested.

Take the hive you drove the bees out of and move it to an enclosed room to harvest its honey and wax.

To whatever extent possible, you should harvest all of the hives that have too much honey—that is, those with almost no empty comb near the bottom. Indeed, such hives would winter less successfully than the others, since bees only cluster well for the winter on the cells containing no honey.

202. Processing honey

Honey is processed to separate the honey itself from wax and pollen. To obtain good honey as simply as possible, use the following method.

In a room—it should be heated up if it isn't warm enough—prepare several large terrines,* with two wicker sieves placed on top. This is where you'll place the comb.

To detach the comb from the hive you're about to harvest, begin by using pliers to remove the sticks inside the hive, then strike it against the ground, first on one side, then on the other. The combs will come loose and collapse atop one another.

---

* *Terrines*—earthenware tureens, usually rectangular in shape, used for baking. Ed.

As you sort through the combs, place newly constructed ones, containing neither brood nor pollen, on the first sieve, and place all the others on the second one.

Use your hand to crush the comb placed on the first sieve; the honey will flow down through it, leaving you with the finest honey in the dish below, referred to as *virgin honey*, which should be poured into a settling tank (§ 129).

Meanwhile, take the comb placed on the second sieve and break it into small pieces without kneading it—then, removing the cappings by scraping them with your thumbnails, open the cells to allow the honey to flow out freely. Before you do, however, be sure to remove all of the pieces of comb containing brood and pollen.

By the next day, most of the honey will have flowed out.

Pour this honey into a separate settling tank than that containing the virgin honey.

To remove the remaining honey that didn't flow out of the old combs, place the sieve holding them atop the bowl (from which you should remove the honey first), then put both into an oven after bread has been baked there. The resulting honey will be of lower quality.

During this process, most of the wax will have melted and flowed through the sieve, forming a layer on the surface of the honey.

You can also use presses to remove the honey from the old combs, but the necessary equipment can be expensive.

To learn more about extracting wax from the other combs, see § 277.

203. Uniting hives after the honey harvest

As we've seen, the bees driven from harvested hives are added to weaker hives that aren't being harvested—but such cases are only frequent in good years.

If the year has been a bad one, then not only might you have no hives to harvest, but you might even have some hives without enough honey for the winter—in which case you'll have to unite these hives amongst themselves.

These hives should be combined shortly after the harvest, so that the newly united bees will have time to organize themselves into a single, unified colony in time for winter.

Any hives with a capacity of approximately 1,250 cubic inches (5 gal, or 20 L) that have less than 13 lb (6 kg) of honey reserves at the end of the season (or less than 26 lb, or 12 kg, for a 2,500-cubic-inch, 10-gal, or 40-L capacity) can't be left as they are.

In a bad year, hives that have less than 4 lb (2 kg) of honey (5-gal, or 20-L capacity) or less than 9 lb or 4 kg (10-gal, or 40-L capacity) often won't be worth the trouble of uniting.

The simplest things to do is drive the bees from these hives, harvest the comb and what little honey it contains—or to leave the hive, with its comb, intact, to house a swarm the following year, after the brood has been removed and the hive has been treated with sulfur.

All of the other hives will need to be united amongst themselves, in pairs, so that each combined pair will have sufficient honey reserves for the winter.

204. How to unite hives after the honey harvest

Let's assume we have two hives following the harvest—one with around 11 lb (5 kg) of honey and the other with 9 lb (4 kg).

After smoking the two colonies, turn each hive upside down and pour some sugar water in between the combs; then, keeping it upside down, place the hive with the least honey in a hole in the ground that you've dug in advance. Next, place the hive with the most honey, right-side up, on top of the upside-down hive (fig. 174, to the left). Seal the gap between the two hives with any appropriate material, leaving only one opening for the entire combined structure.

Use wire hooks to attach a properly sized piece of comb to one of the largest combs of the upper hive, so that this piece of comb touches another comb of the lower hive; this added piece of comb will serve as a kind of passage for the bees between the two hives.

The two combined colonies will eliminate one of the queens. Generally, the honey in the lower, upside-down hive will be consumed first, or be moved into the upper hive.

Ideally, you should unite two hives that are nearby one another so that the bees from the hive that was combined with a stronger hive will return more easily to the location of the united hives, despite not

Figure 174. Uniting hives using the superposition method. The hive to the right has not been combined.

finding their old hive when they return as usual to their old, accustomed spot. If you unite hives that were located far away from each other, you run the risk of losing bees.

Generally, this procedure is quite successful. However, the following exceptional situations may arise:
1. Both queens are killed—an extremely rare occurrence.
2. The colony may install itself in the lower hive only, which you can confirm in the spring when you find brood in the lower hive only. In this case, remove the upper hive and turn the lower one right-side up; the hives will merge in the opposite direction.
3. Sometimes, you may find bees in both hives in the spring, but brood in only one of them—in which case you should keep the hive with the brood and remove the other.

Figure 175. A beekeeper placing a dish full of syrup underneath a hive to feed the bees.

205. When fall feeding is required

After a poor season, all of your hives may lack sufficient honey reserves to survive the winter—or even to be combined amongst themselves. In this case, you'll need to feed half of the hives (fig. 175) that have the most bees; then, when the feeding is complete, unite the hives you've fed with others, in pairs.

Here is what our model apiary may look like after an average season—if, for example, five hives were completely harvested (or sold). This leaves:

1. Five hives, to which the five harvested strong hives have been completely added.
2. Eight hives, with sufficient winter reserves, left unaltered.
3. One hive, low on honey, that has been fed.
4. One hive resulting from uniting two hives that had insufficient reserves.

Figure 176. A fixed-comb hive during wintering. You can see the mouse guard covering the entrance and the two wedges propping up the hive slightly above its stand.

Based on these assumptions, fifteen hives in good condition will be entering the winter period, like last year, with five of the strongest hives having been harvested or sold.

206. Readying for winter

You can ready fixed-comb hives for winter as described previously (§ 76). As for the hives placed atop one another while being united, you can simply use some perforated sheet metal or wires, arranged as necessary where the two hives meet, to prevent field mice from getting in.

207. Managing cap hives

If you're using cap hives as described in § 44, here's how to manage them.

At the end of the wintering period, the hives lack caps, which should be added a few days before the main honey flow begins.

To add a cap (fig. 177), unseal the hole at the top of the hive and set the cap in place, using putty to seal the junction around its entire perimeter, if necessary. Inside the cap, attach a long piece of comb whose bottom will touch the hive's upper combs through the hole, to ease the bees' passage between the hive and cap.

If the honey flow is very strong, remove the cap when it's full and replace it with an empty one.

To remove a cap (fig. 178), lift it up, blowing smoke into the gap,

CHAPTER 15. TRADITIONAL FIXED-COMB HIVES 231

Figure 177. Adding a cap to a hive.

Figure 178. Removing a cap.

and take it to your honey room to drive the bees from it—then, follow the procedure used for the supers in § 188.

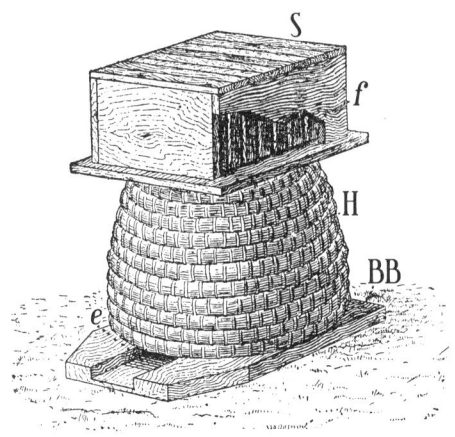

Figure 179. A fixed-comb hive made of woven straw, with a frame super. *BB*, bottom board; *e*, entrance; *H*, hive body; *S*, frame super; *f*, frames, visible here through a partial cross-section of one wall.

208. Frame caps

Instead of regular caps, you have the option of adding *frame caps* to fixed-comb hives. You can cut off the top of a traditional fixed-comb hive and attach a board with open slits cut into it. The strips of wood left between the slits should be primed with bits of comb to guide the bees as they build comb in the body of the fixed-comb hive.

The frame cap can be placed atop this board with slits cut into it; each frame will line up with the interval between two slits.

These hives can be managed like vertical hives with supers. In winter, replace the frame cap with a board.

You can also replace frame caps with section boxes (fig. 162).

Summary

Managing fixed-comb hives

If you plan to continue using nothing but traditional fixed-comb hives for several years or more, you can manage them as follows.

Once the wintering period is over, conduct your spring hive inspection. Eliminate dead and disorganized hives. Feed any strong hives that are low on honey, and, during honey flow season, swap the locations of weak hives that have wintered successfully with strong ones.

If you allow natural swarming to occur, collect the prime swarms, return any afterswarms to their hives of origin, and unite any weak or late swarms.

CHAPTER 15. TRADITIONAL FIXED-COMB HIVES

Artificial swarming and fixed-comb hives

If you'd like to reduce natural swarming and avoid combining swarms, you can pursue artificial swarming, which has the advantage of producing swarms even in years when they would not emerge naturally.

One of the best ways to produce an artificial swarm involves two strong hives that are relocated; this procedure results, by season's end, in three hives in place of two.

Harvesting fixed-comb hives, and wintering

Some hives are harvested or sold. The best way to harvest fixed-comb hives is to completely harvest a certain number of strong hives from which the bees have been driven, to be added to other, weaker hives.

The harvested honey must be processed—that is, separated from the wax and pollen. The wax is then melted.

In bad years, hives with insufficient winter reserves must be united in pairs following the harvest.

During particularly bad years, you'll need to feed half of the hives before uniting them with others.

Cap hives

Generally speaking, cap hives are managed in the same way as vertical hives with supers. If you like, you can replace caps with frame boxes.

Notes on fixed-comb hives

All things considered, using fixed-comb hives has the advantage of requiring very little money, and in any event can be recommended to beekeepers interested in breeding—that is, selling populated hives. On the other hand, managing fixed-comb hives successfully requires more time and labor than any other commonly used hive variety—not to mention the fact that they typically require a much more experienced keeper than vertical frame hives in order to use them to their full potential.

Chapter 16
Supplementary Material

209. General observations

Up to this point, we've focused on the kinds of hives that we believe to be the most practical for managing bees according to various methods; we've also made use of a variety of beekeeping tools that are more than sufficient for carrying out all of the necessary procedures.

However, there are many other hive varieties, and your set of beekeeping equipment can be expanded to include tools that will further simplify the procedures we've discussed, or that are involved in other supplementary techniques to be covered in the following chapter.

Still, it must be said that improvements made to a tool may prove, in the end, to be little more than an unnecessary complication—and although a new invention may represent a step forward in some sense, using it may also give rise to new problems. We cannot improve beekeeping by going farther and farther away from the bees' natural tendencies.

There is a very simple way to judge the usefulness of a proposed innovation: make a numerical estimate of the extent to which its use will increase your annual harvest, while taking into account the additional expenses and time it will require.

If a given innovation won't increase the harvest, or might even lower it, then it could only be recommended for those who view beekeeping as a hobby, not as a for-profit enterprise.

210. Frame hives similar to those already described

In order to keep things simple for beginners, our discussion up to now has focused on one system of horizontal hive, with a given set of dimensions. We did the same for vertical hives.

People have used or proposed hives that are similar to the ones we've chosen, or whose only major difference is the dimensions of the frames.

1. *Hives with frames that are deeper than they are long.* The moveable-frame hive we've described—the so-called French (Layens) hive—has frames that are deeper than they are long; this frame shape is meant to ease wintering: during a long cold spell, it allows the bees to move upward along the frame over a longer period, consuming the honey contained therein as they go, without breaking up their cluster.

There are other, very similar hives whose frames, although still deeper than long, have slightly different interior dimensions, such as: 15 3/4" (40 cm) deep and 11 5/8" (30 cm) long.

Only rarely has anyone proposed making *vertical* movable-frame hives with frames that are deeper than they are long, since, as we've seen, this shape makes it difficult for the bees to make their way up into the supers.

2. *Hives with frames that are longer than they are deep.* The vertical moveable-frame hive we've described has frames that are longer than they are deep—and the supers of such hives have frames that are even shallower than those in the hive body.

There are very similar hives whose frames, although still longer than deep, have slightly different interior dimensions, such as: 8 1/4" (21 cm) deep and 17" (43 cm) long; 10 5/8" (27 cm) deep and 18 1/8" (46 cm) long; 11 5/8" (30 cm) deep and 15 3/4" (40 cm) long, etc.

Only rarely are horizontal hives made with frames that are longer than deep.

3. *Hives with square frames.* There are also hives that fall somewhere in between the two frame shapes just described: the depth and length of these frames are equal.

There are also horizontal hives that have square frames.

The interior dimension of one side of these squares can vary between 13 3/4", 13", 12 3/4", 11 1/4" (35, 33, 32.5, 28.5 cm), etc.

211. Various systems of moveable-frame hive

So many other varieties of hive have been invented that it would be difficult to list all of them. Suffice it to mention several kinds of horizontal hives, vertical hives, and so-called "warm-way" hives.

212. Varieties of horizontal hives

The *leaf hive*, invented by Huber, is a hive consisting of nothing but frames pressed tightly together and tied, with no external casing. The two outermost frames are made of solid wood.

Various beekeepers have tried various modifications of the leaf hive—for example, giving the frames a rounded shape at the top.

Without dwelling on other kinds of horizontal hives consisting of frames inside a hive body—like those proposed by Blake, Munn, Prokopovich and de Beauvoys—we might mention the *German horizontal hive*, invented by Gravenhorst (fig. 180), which consists of frames rounded at the top, as just mentioned, inside a casing of woven straw, and removable from the bottom only.

There are numerous varieties of horizontal hives with frames inside

Figure 180. A Gravenhorst horizontal moveable-frame hive.

a hive body. In essence, they all have the same structure as the hive we adopted: a series of frames, in a single row, removable from the top.

These horizontal hives are sometimes referred to as *long hives*, since they usually contain enough frames to be elongated in the direction perpendicular to the combs. These include hives invented by Warquin, Thierry-Mieg, Santonax, Sagot (with frames topped with triangular frames, fig. 181), Brunet (with circular frames), etc.

We should also mention the *album hive* invented by Charles Derosne: a horizontal hive with an ingenious mechanism for swiveling the frames around their end bars when inspecting the hive.

213.1. Varieties of vertical hives

There are also a large number of vertical hives.

Hives whose supers contain sections or frames that are shallower than the brood frames in the hive body include, alongside the Langstroth hive, the *Quinby hive* and the *Adair hive*; those with square frames include the *Gallup hive* (11 1/4", or 28.5 cm, on each side), the *Voirnot hive* (13", or 33 cm, on each side), and the *standard American hive* (12", or 30.8 cm, on each side).

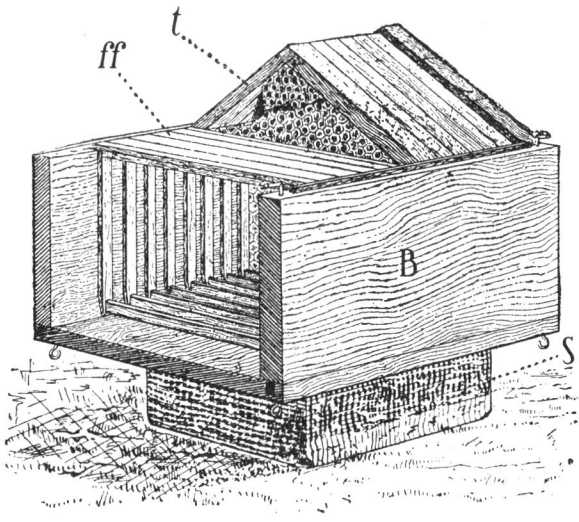

Figure 181. A Sagot moveable-frame hive. *S*, stand; *B*, hive body; *ff*, frames; *t*, triangular surplus frames placed on top of the hive.

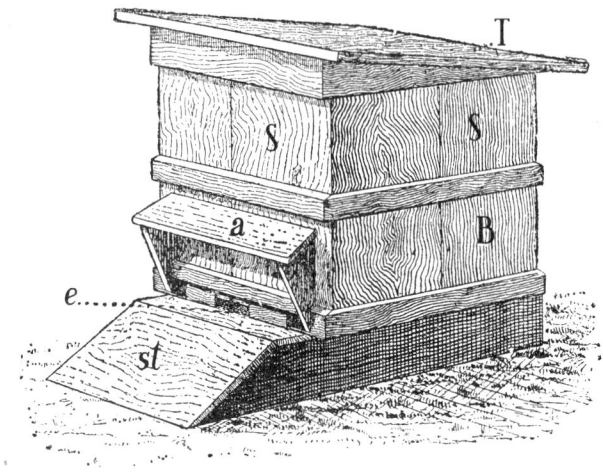

Figure 182. An English vertical hive (shown here with a single super). *B*, hive body; *st*, stand; *e*, entrance; *a*, entrance cover; *S*, super; *T*, top.

English hives (fig. 182) are also small vertical hives designed specifically for producing section honey.

The standard frame used in England has interior dimensions of 8" (20.3 cm) in depth and 13 1/2" (34.3 cm) in length.

Other vertical hives consist of multiple, identical hive bodies that are stacked on one another; these include the *Root hive*.

213.2. Hives employing a two-queen system

The idea of housing two colonies working side by side, with a common cap or super, is an old one; its goal is to achieve a strong population during foraging season, consisting of two strong colonies working together. This arrangement was meant to boost the overall honey harvest.

Jules Devauchelle applied this system to moveable-frame hives; George Wells has recently modified it.

The *Wells hive* basically consists of a horizontal hive completely divided in the middle so as to house two colonies, to the left and right of this division. The board separating the two hives has a grating at the top of around 8" (20 cm) on each side that does not allow the bees to pass, but does allow for the two colonies to have a single, uniform odor. During foraging season, a single, shared super is placed atop the structure, providing communication between the two colonies at the top; on top of the frames inside the two hives themselves, a perforated sheet of metal is placed that allows workers to pass, but not queens.

214. "Warm-way" hives

All of the hives we've discussed so far allow you to remove any given frame without moving the others—either from the bottom, from the side, or, more often, from the top. Since the entrance of these hives is positioned perpendicularly to the combs, multiple gaps in between combs are ventilated simultaneously. All such hive systems can be grouped into one general category: *"cold-way" hives.*

There is another system of moveable-frame hives, both horizontal and vertical, where the entrance is positioned parallel to the combs. This means that the outside air only hits the nearest comb directly, before passing successively into all of the gaps between the combs.

Such hives can be grouped under a second general category: that of *"warm-way" hives.*

Figure 183. Three-compartment German hive with "warm-way" frames.

With this hive system (fig. 183), the frames are removed by their front, not their side, which means that if you need to inspect the last frame, you'll have to remove all of the others. You can't simply remove the frames through the top.

There are also stand-alone warm-way hives, like those we've described throughout the book; but this hive variety is typically used as follows.

All of the hives are placed side by side inside a kind of shed, positioned with the entrances facing outdoors, allowing them to be opened on the opposite side—that is, from inside the shelter, much as one might open a cupboard.

This allows the beekeeper to work while remaining protected from the wind and rain, and with less chance of being stung; this setup also saves the trouble of moving all of the necessary equipment from one hive to another.

But this arrangement also involves significant problems: it's impossible to change the positions of the hives during certain procedures; inspecting each hive is a long and complicated process; and, if foulbrood strikes a single colony, it can spread throughout the beehouse and destroy the entire apiary.

Due to long-standing custom, this beehouse system remains the most common in Germany and in certain neighboring countries.

There are many varieties of warm-way hives; the main ones include those of Dzierzon, Berlepsch, Bastian, Sartori, Burki, Jeker, etc.

215. Varieties of fixed-comb hives

There are also numerous models of fixed-comb hives. We've already discussed (beginning in § 42) log hives, hives made of boards, bell-shaped hives (skeps), cap hives, and stacked hives.

Let's look at a few modifications of these various hive systems.

In Algeria—and, generally speaking, in Middle Eastern and Asian countries—a very shallow hive is used, made of resinous wood or of the stems of the fennel plant.* These hives are of the horizontal fixed-comb variety.

---

* Giant fennel (*Ferula communis*) is a herbaceous perennial plant in the Umbelliferae (carrot) family whose hollow stems have a strong odor.

Chapter 16. Supplementary Material

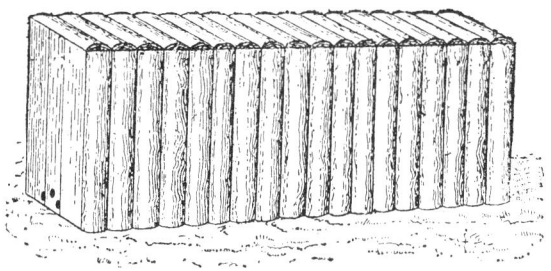

Figure 184. Arabic hive.

Figure 185. Corsican hive.

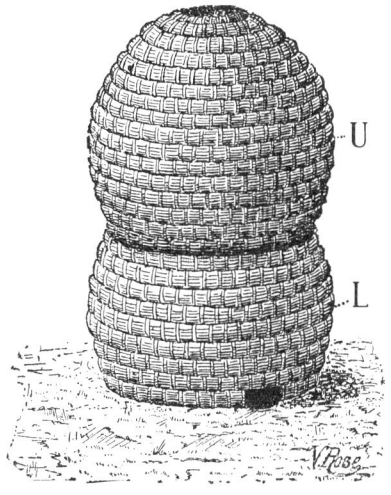

Figure 186. Scotch hive. *L*, lower section with entrance; *U*, upper section.

The same is true in Corsica, but there the hive consists of a log laid horizontally (fig. 185).

In Egypt, these shallow, horizontal hives are made of terracotta.

The *Scotch hive* is a cap hive whose cap is as large as the hive body. This large cap can be placed on top, like a super, or, sometimes, underneath, when the combs in the hive itself need to be replaced.

The *Lombard hive* is a hive whose cap is put back on top of the hive during winter, following the harvest.

There are also such things as wooden stacked hives with guides on top to force the bees to build their comb regularly.

216. Choosing a hive

We've just listed many kinds of hives, falling under many different hive systems. So what hive should a beekeeper choose?

The hives we've described in detail in the preceding chapters were chosen from among time-tested models that are now the ones most commonly used in our country.

If you adopt these hives, or other models that differ slightly, you can expect to maximize your yield, for your given region, while minimizing the time and expenses required.

But, of course, you're free to pursue another system that may be more expensive or more complicated.

A good keeper will manage to extract the best possible returns from any system of hive, since, when it comes to beekeeping, a deep knowledge of the bees' way of life is always more important than the kind of housing you choose to provide for them.

217. Observation hive

You may be interested to watch bees at work inside a hive. This is precisely why hives called *observation hives* are sometimes made (fig. 187).

The most convenient kind of observation hive simply consists of a single frame, closed on each side by a pane of glass, covered by a shutter. By opening these shutters on one side or the other, you can observe the bees as they go about their work inside this tiny colony.

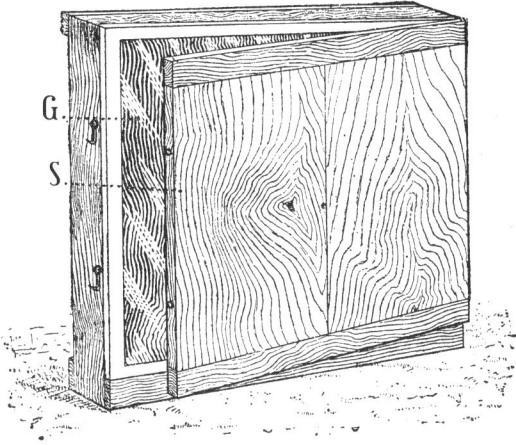

Figure 187. Observation hive. *S*, shutter; *G*, glass pane allowing for observation of the bees at work.

You can watch the queen laying her eggs, and the workers storing honey and pollen, building queen cells, etc.

To populate such a hive, you can take a frame with brood of all ages from a strong colony, along with any bees that happen to be on it, but without the queen. Since the oldest bees will for the most part return to their old hive, and since the bees left behind on the frame will be too few to maintain sufficient warmth for the brood, you should take the following measures.

Take another comb, similar to the first—along with its bees, but without the queen—and sweep them off the comb in front of the observation hive's entrance.

Many of these bees will enter the hive, providing a sufficient boost to its population.

To go one step further in ensuring that as many bees as possible remain in this observation hive, place it in a cellar for forty-eight hours, having covered the entrance with a wire mesh fine enough to keep the bees from getting through.

218. Advantages and disadvantages of a beehouse

If space is tight, but you'd still like to have a good number of colonies, you can arrange your hives in tiers, in what is called a *beehouse*.

The hives are usually set up in two tiers, and protected from the weather by some kind of overhead shelter.

A beehouse has a few advantages: it takes up little space; your tools will always be close at hand during any procedures; and the hives themselves will have no need for special roofing to protect them from the rain.

But these advantages of this arrangement are far outweighed by its problems.

Obviously, many procedures become difficult when dealing with the hives on the upper tier—especially when you're required to move hives.

But there's another, more serious problem: when young queens return to the apiary after their mating flight, they'll often enter the wrong hive, despite any precautions you may take to differentiate the hives from each other, whether by painting them different colors or by using some other device. The result is a larger number of orphaned hives.

All things considered, the use of beehouses (which, by the way, are only found in a few regions of France) can't be recommended.

219. Scales, thermometers, hygrometers, barometers, and microscopes

Generally speaking, any instrument used in meteorology can be of use in beekeeping.

In any larger apiary, it's useful to keep one or several hives on *scales* to keep track of how foraging is progressing.

Note, however, that the readings on the scales aren't always absolute: fluctuations in the number of bees, the quantity of brood, the weight of gathered pollen or water can cause significant variations in the hive's weight, independent of the weight of honey produced. But since during the foraging season the weight of the honey is the predominant factor in a hive's overall weight, the scales can provide their most useful readings during a heavy honey flow. Still, keep in mind that the nectar gathered by the bees contains a lot more water than capped honey—and this loss of water, which evaporates in large quantities thanks to the ventilation provided by the bees, accounts for an apparent drop in the weight of the hive's honey, especially at night.

CHAPTER 16. SUPPLEMENTARY MATERIAL

The best way to get a rough idea of how foraging is going is to take one reading per day, in the evening, once the bees have returned to the hive—always taking readings at the same time each day.

A *thermometer* can tell the beekeeper whether certain procedures should be carried out or not, whether swarms are likely to emerge, etc.

Barometers and *hygrometers*, together with wind direction and, above all, knowledge of meteorological indicators specific to your region, can help predict the weather—but when it comes to weather predictions, beekeepers are often wrong, and, of course, we can never know for certain what the weather will be. This is why it's unwise to use certain beekeeping methods (see § 231, for example) that require you to predict the weather.

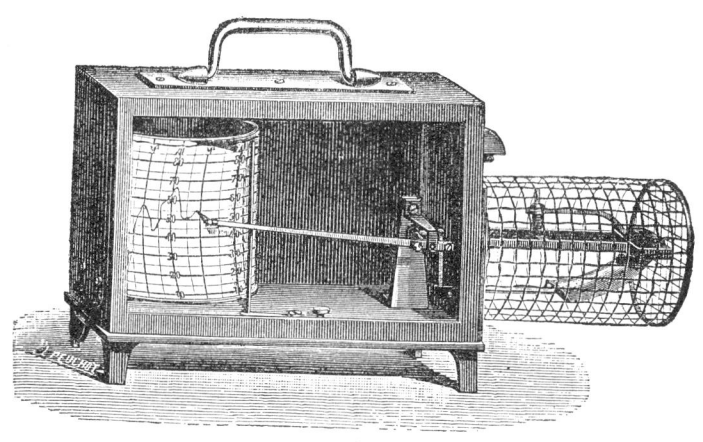

Figure 188. Recording thermometer.

The instruments we've just mentioned can be replaced by recording instruments. So a recording thermometer as shown in figure 188 can continuously record the temperature for eight days, without being reset. Such instruments can be placed inside a hive.

A *microscope* can open up an enormous wealth of scientific observations for an amateur beekeeper—for example, examining foulbrood bacteria, the highly varied shapes of pollen from various plants, or the yeast that causes mead to ferment, the nectar glands of flowers, bee anatomy, etc.

220. Feeders. Feeder varieties

We've described how to feed hives using a simple bowl, or a mason jar turned upside-down.

Now, we'll look at a very simple device, called a *feeder*, designed for use with moveable-frame hives.

Figure 189 depicts such a feeder. It's a tinplate box into which syrup is poured through an opening, *o*, closed by a plug, *pl*. This box is placed upside-down above the gap between two frames after the slat has been removed; the parts of the gap left uncovered by the box are plugged using pieces of wood.* In this position, the feeder presents itself to the bees as a flat surface, *perf*, perforated with holes too small to allow the honey to flow out, but large enough to allow the bees to come and gather the sugary liquid.

It's best to carry out this procedure in the late afternoon; the next morning the feeder will be empty.

There are other kinds of feeders as well.

The simplest is an upside-down bottle. A bottle filled with syrup is placed inside a hive, and turned upside down in small tinplate trough; the bottle is inclined to varying degrees so that the syrup in the trough is replenished at the same rate by which the bees consume it.

A tinplate vase, plugged by a piece of light fabric and turned upside-down over the hive, can also serve as a feeder.

We might also mention English feeders such as that shown in figure 191, the Raynor feeder, which allows you to regulate the syrup flow (fig. 190), as well as the Derosne feeder (fig. 192).

221. A water-bath oil can

You can use a special kind of oil can to prime frames with a bead of wax (§ 102), or to affix the top of foundation sheet to a frame (§ 99).

This oil can (fig. 193) allows you to keep wax melted using a hot-water bath inside the can.

---

* If you have a hive with cover boards on top instead of slats between top bars (see note to § 98), place the feeder upside-down in a square hole cut into one of the cover boards located above the frames.

CHAPTER 16. SUPPLEMENTARY MATERIAL

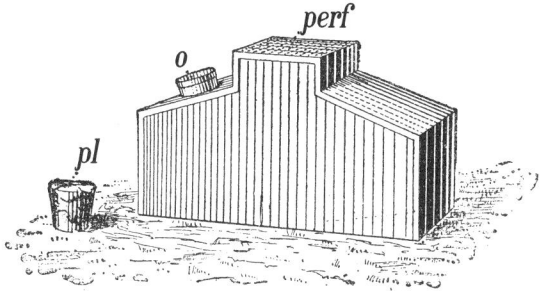

Figure 189. A Layens feeder in which syrup has just been poured. During use, the feeder is turned upside-down and placed atop a moveable-frame hive: *o*, opening through which syrup is poured, sealed using a plug, *pl*; *perf*, the side perforated by small holes where the bees come to collect the syrup.

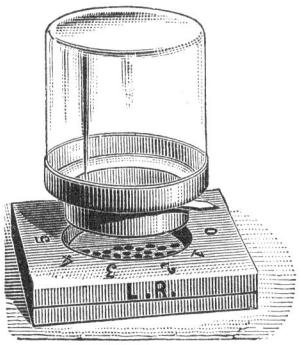

Figure 190. A simplified Raynor feeder. By rotating the feeder, you can adjust the amount of syrup dispensed.

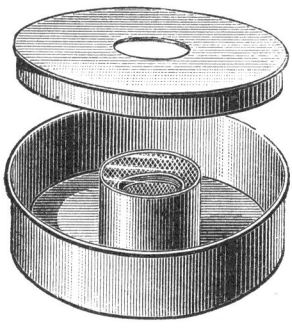

Figure 191. An English feeder made of tin with interior cylinder.

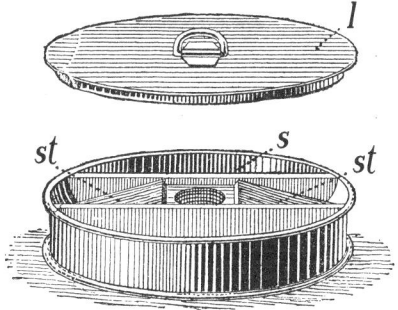

Figure 192. A Derosne feeder in the position it should occupy inside the hive: *s*, one of two lateral sections into which syrup is poured; *st*, strips where the bees come to feed; *l*, lid.

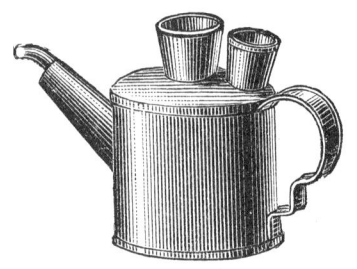

Figure 193. A burette that keeps the wax melted inside a hot water bath.

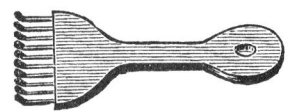

Figure 194. An uncapping fork.

222. An uncapping fork

Figure 194 shows a small tool called an *uncapping fork*, designed to tear open caps. But this procedure is more easily carried out using a knife of some sort.

223. Bee repellents

Various liquid sprays have been invented, called *bee repellents* (their ingredients are kept secret by the manufacturers), which you can apply to your hands in order to avoid being stung, even while working without gloves. Some bee repellents may present a health hazard.

Be that as it may, keep in mind that if you yourself are completely protected from bee stings, and work without the usual precautions, then others around you may still be at risk for being stung—that is, you'll be less worried about agitating the bees, and they'll look for someone else nearby to sting.

224. Drone traps

Many kinds of traps have been invented to get rid of drones when there are too many of them.*

The devices used for this purpose are called *drone traps.*

---

* As previously noted, having a large number of drones contributes to the genetic success of the colony and the overall health of the honey bee population. We now know that the queen needs a dozen drones or more (from colonies other than her own) for optimal mating, and that the diversity of traits inherited from the different drones is very important for the functioning of the colony. Following the natural blueprint, many beekeepers today do not try to limit the number of drones a colony produces. *Ed.*

Figure 195 depicts one such device.

Like almost any other form of trap, this one is based on the use of perforated metal. This metal is perforated with rectangular holes that are 1/2" (13 mm) by 3/16" (4 mm). Only worker bees can make it through these openings.

Figure 195 shows a box (whose sides are made of such perforated metal) placed in front of the hive entrance. In between the area at the bottom of the box, shaded in gray, and that at the top (in white), there are special valves that allow drones to pass into the upper section of the box, but not to re-enter the hive, while the worker bees can go anywhere they like, through the perforated metal. The drones begin to add up in the trap—when it's sufficiently full, remove the box and throw the drones in water.

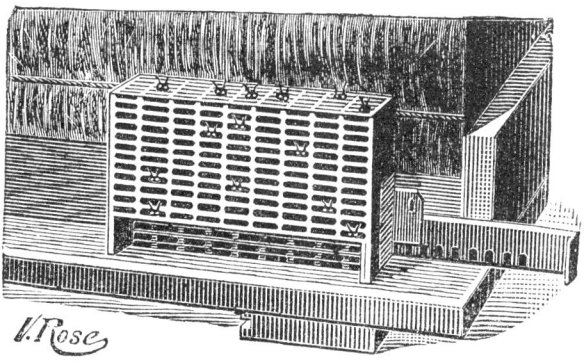

Figure 195. A drone trap.

You can get rid of the vast majority of your drones in just a few days using this technique.

But you shouldn't leave this trap deployed for too long—if the colony happens to be replacing its queen, the drone trap will prevent the virgin queen from departing on her mating flight.

Nor should you deploy the drone trap during the main honey flow, since this added obstacle will disturb the bees, who will waste valuable foraging time learning to navigate it.

225. Varieties of extractors

1. *The reversible-frame extractor.* There are extractors whose exterior resembles that shown in figure 47, but which have the advantage of always preventing combs from breaking, and allow beekeepers to extract thick honey—even at low temperatures—without breaking combs.

The inside of the extractor is made of steel, with two baskets, into which the uncapped combs are placed; the stiffness of the steel bars supporting the grating prevents it from bending and breaking the comb.

Each of these baskets is reversible—that is, it can swivel so that the side previously facing inward is turned outward.

So after extracting most of the honey from the outer side of the two frames, you can swivel each frame 180° to completely empty the other side—before finally swiveling the frames once again to complete the extraction for the first two sides.

2. *Budget extractors.* Various kinds of less expensive extractors have been invented. They usually consist of a wooden barrel with a kind of reel in the center, surrounded by cords that do the work of the metal grating in ordinary extractors. These cords have the advantage of simply tightening instead of bending outward when the extractor is in motion.

226. Bee escapes

Various tools have been devised to drive bees from a section of a hive in order to simplify the harvest.

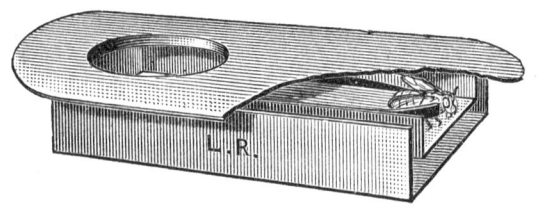

Figure 196. A bee escape.

One such tool is called a *bee escape* (fig. 196). When a bee approaches from above, it passes through a hole in the top; in order to exit, it must push aside two lightweight elastic tabs that then close shut behind the bee, preventing it from re-entering.

A board with an embedded bee escape—this tinplate trap that will only allow the bees to pass through the board in a single direction—can be placed below the super of a vertical hive, or in place of a frame in a horizontal hive.

When the bees move to the center of the colony, they'll be unable to climb back up into the super, or into the far end of a horizontal hive—the area on the other side of the board.

However, before installing a bee escape, you should inspect the supers to make sure they don't contain any brood—or, you should inspect a horizontal hive in order to determine which frame should be replaced by the board.

If, for example, a hive has three supers, remove them one by one; then, with an assistant, fit the bee escape atop the hive body and put the supers back into position. Wait for one day before removing the supers once again for harvesting, which will now be a simpler task, provided that the bees have been kind enough to pass through the bee escape.

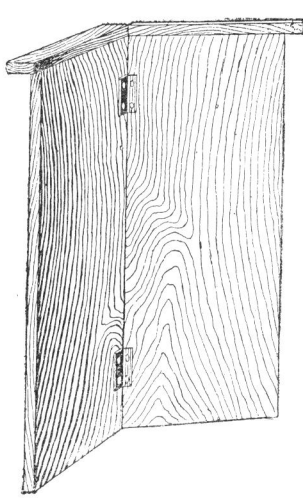

Figure 197. A division board.

However, this simplification is only a matter of appearance. We've already described (in § 188) a quicker method that doesn't involve a bee escape.

227. Division boards

Figure 197 shows what is called a *division board*, used to reduce the volume of the hive body during certain procedures. Such partitions were used in the past, primarily to concentrate a colony's warmth during winter.

But careful experimentation has since demonstrated that a built-out frame serves precisely the same purpose, rendering division boards completely useless.*

Summary

Varieties of hives

There are hives that differ from those described in detail earlier in the book only in terms of frame dimensions; most beekeepers believe that a good moveable-frame hive should have frames with a surface area of around 186 square inches (1,200 cm^2).

Many models of horizontal and vertical hives have been proposed or actually built; most often, they feature smaller frames than those just mentioned.

Moveable-frame hives from which any given frame can be removed without moving the others are called *cold-way hives*.

Germany and certain neighboring countries use moveable-frame hives whose top is fixed in place; removing the last frame requires removing all the others. These are called *warm-way hives*.

There are also numerous modifications of fixed-comb hives.

All in all, the hives we've chosen for the various beekeeping systems explored in this book are those that appear to be both the simplest and the most practical.

Supplementary beekeeping equipment

The beekeeping equipment commonly used by an amateur can be supplemented with the following: an observation hive, for an inside view of how bees work; scales, for keeping track of how the foraging is progressing; a thermometer, hygrometer and barometer for following changes in the weather; a microscope, for various kinds of scientific study; and, finally, bee repellents, drone traps, and bee escapes.

In certain regions, hives are installed in a beehouse, which presents more problems than benefits.

There are also various kinds of feeders and extractors. Among the latter, the steel reversible-frame extractor is one of the best.

---

* For more details, see: G. Bonnier, *Expériences sur l'inutilité des planches de partition*, G. de Layens, *Nouvelles expériences pratiques d'apiculture*, p. 17. (See also chapter 24. Division boards may be of use when a newly installed swarm is drawing comb, especially on foundationless frames. Many northern beekeepers find them useful in winter as well. *Ed.*)

Chapter 17
Alternative Procedures

228. General observations

In previous chapters, for each beekeeping system, each with its own special goals, we've chosen to use those procedures that seem the simplest and the most practical.

However, there are other methods out there, which, in certain specific instances, may be substituted for the procedures we've described. Some of these alternative procedures are almost just as simple and as effective as those seen thus far, and, as a beekeeper, you're certainly free to adopt them. Others, however, are more complicated or uncertain—in which case we'll point out their drawbacks, alongside the benefits they can provide.

So let's examine the various procedures we've left out of our preceding discussion, organized, as much as possible, in their natural order.

229. Buying populated moveable-frame hives

The ability to buy moveable-frame hives already containing strong, fully installed colonies would make things considerably easier for a beginning beekeeper—by saving you the trouble of buying fixed-comb hives, wintering them, and transferring their colonies into moveable-frame hives. On top of that, you'd be certain to have a harvest in year one, provided the year is a normal one.

Unfortunately, it's still rare to find fully populated moveable-frame hives for sale. If you do find them, you should only buy them if they are very strong—in the spring, for example, when the willow is blooming, the hive should have brood on at least four frames, and should be filled with fully built-out frames.

In any case, a beginner who is unable to find populated moveable-frame hives for sale will be compelled, as we've seen, to practice handling bees using fixed-comb hives—and this experience will prove highly valuable down the road.

230. Transfer by superposition or by artificial swarming

We've discussed multiple transfer methods (beginning in § 142). Here are two more.

1. *Transfer by superposition.* This procedure is especially suited for low-capacity fixed-comb hives. It essentially involves placing the fixed-comb hive atop the prepared moveable-frame hive, and letting the bees make their way downward over the course of the foraging season. One good way to encourage the bees to pass into the moveable-frame hive is to further reduce the volume of the fixed-comb hive.

Trim all of the combs along the bottom, up to the point where the brood is located, then saw off the entire perimeter of the hive up to the level the comb was trimmed to. Now that the hive has been cut off at the bottom, it can be placed atop the frames of the empty hive that now stands in the location previously occupied by the fixed-comb hive. Seal up the

Figure 198. Superposition transfer from a fixed-comb hive into a moveable-frame hive. The fixed-comb hive is placed on top of the moveable-frame hive, and the entirety is covered with a straw hackle.

parts of the moveable-frame hive that aren't covered with the fixed-comb hive using some rags and boards, so that the bees will have to pass through the moveable-frame hive if they wish to enter or exit. Cover the entire structure with a large straw hackle, to protect both hives at once (fig. 198).

This procedure should be carried out in early spring.

If the year has been a bad one, the bees may not climb down into the moveable-frame hive. In this case, leave the hives as they are throughout the winter, and wait for next year.

2. *Transfer by artificial swarm.* Select two strong fixed-comb hives, A and B, and drive the bees from hive B into an empty fixed-comb hive, from which they will be made to pass into a moveable-frame hive prepared especially for this purpose. Once the bees are inside this moveable-frame hive, place it in the former location of hive B, then place hive B in the former location of hive A, which should be relocated a good distance away. Twenty-one days later, hive B no longer contains any brood; harvest it completely, after adding its bees to one of the weak hives in your apiary. In this fashion, you will have transformed a fixed-comb hive into a moveable-frame hive.

231. Speculative feeding

From time to time, someone suggests feeding hives in small, successive doses during certain times of the year—and not because these hives are low on honey reserves, but to make the bees believe that nectar is being gathered even though there's no honey flow outside yet, thus artificially stimulating the queen's egg-laying.

This artificial method for promoting egg-laying is meant to generate a higher population in advance of the actual honey flow. It's referred to as *stimulative feeding* or *speculative feeding*. The term "speculative" is an apt one, since, as we'll see, this procedure has all of the drawbacks one would associate with any kind of speculation.

This procedure can be useful, useless, or even harmful. Since it's generally impossible to predict in what conditions speculative feeding will prove good or bad, and since we can't foretell the weather, this approach is quite a gamble.

If you anticipate that the main honey flow will take place during a certain period, then six or seven weeks before this period begins you can start feeding each hive small but increasing doses of syrup; the doses can range from 2 to 9 oz (50–250 g). The feeding should never be interrupted, since the bees will need more and more sugar as the egg-laying intensifies. This daily distribution of syrup should be carried out with every possible precaution, to prevent robbing. You can well imagine the stress and labor required by such an extended procedure.

If speculative feeding is to succeed, the following must hold true:

1. *You must be certain of the dates of the main honey flow—and that it will even happen at all.* In fact, if your predictions are wrong, all of the time and sugar you expend will have been wasted. To make matters worse, you will have built up a large but useless population that—in the worst-case scenario of zero honey flow—will consume existing honey reserves in even greater amounts.

2. *There must be no sudden daytime drop in temperature while speculative feeding is underway.* If, as quite frequently happens, the temperature drops too much, once or repeatedly during feeding, the bees—agitated by the artificial honey flow you've provided—will exit the hive in the belief that they're embarking on a foraging flight, only to fall numb from the cold. In this case, speculative feeding may actually reduce the number of bees instead of raising it—and you will have wasted your time and money on a negative outcome.

3. *There must be no intense and extended cold period lasting several days during the feeding.*

If this happens—and it can never be foreseen—the bees will be forced to cluster, and will have to abandon any part of the brood they're no longer able to cover. And this abandoned brood may develop foulbrood that could lead to a loss of hives.

If these dire circumstances come about, you will have wasted time and money—and you yourself will have brought about the ruin of your apiary.*

---

* Conclusive experiments have shown that speculative feeding is at least useless, and at worst problematic. (See *Nourissement stimulant*, by abbot Martin, former president of the Société d'Apiculture de l'Est, in *Apiculteur*, 1890, p. 193.)

All things considered, speculative feeding can only be recommended in those happy regions where long-term meteorological observations give complete assurance that the three preceding conditions will always be met.

In any case, you can take one risk-free step when arranging the frames in the spring as we recommended previously. By uncapping the honey cells located above the brood, you will not only give more room to the queen, as we mentioned, but you'll also stimulate egg-laying early in the season to an extent that poses no problems, and with no additional time or labor required.

232. Feeding with sugar paste

If you're compelled to feed your hives at a time of year when it's no longer appropriate to do so by giving bees combs filled with syrup, and when you've run out of reserve frames, you can use the following method.

Make a paste using 1 qt of honey and 7 1/4 lb of powdered sugar (2 lb honey, 5 lb sugar; 1 L honey, 3.5 kg sugar). First, heat up the honey, then mix the powdered sugar into it, mixing thoroughly. Take the desired amount of paste, wrap it in a light cloth, and place it—spreading out the cloth—on top of the frames, above the cluster of bees, then cover the hive back up as usual.

233. Other methods for preventing afterswarms

We've provided one method for discouraging afterswarms (§ 113). Here are two others.

1. *Displacement.* When, after the primary swarm has already emerged, or after an artificial swarm has been formed, you hear the queen's song, move the parent hive to another location more than 16 ft (5 m) away. Now, most of the bees that leave the hive for foraging will end up in other hives, and this drop in population will prevent the afterswarm from developing.

As you can see, this method is extremely simple.

2. *Removal of queen cells.* Sometimes, you can prevent afterswarms by removing all the queen cells, except for one, from the parent hive.

This method has one disadvantage: the single queen cell you leave in place could contain a deficient queen, or even a dead larva. If this proves to be the case, then you will have singlehandedly orphaned the hive.

234. Artificial swarming with one hive*

We've already provided a sound method for producing artificial swarms—one that requires two strong hives, and is equally appropriate for fixed-comb hives (§ 200) and moveable-frame hives (§ 163). Here's another, which works only for moveable-frame hives, and involves only a single hive—but is not nearly as good or dependable as the method previously described.

Figure 199. *A*, a strong hive from which two brood frames are removed, along with the bees covering them, and put into an empty hive, *B*.

Figure 200. *B*, having received the two brood frames (with bees) from hive *A*, as well as a honey frame and five or six built-out but empty frames; this hive *B* is put in the former location of *A*, and welcomes any bees who were away foraging. *A*, the relocated parent hive.

On a nice day, shortly before the start of swarming season, when the bees are highly active, take two brood frames, along with any bees on them—one of the frames should contain brood of all ages—from a thriving hive (*A*, fig. 199) with at least seven brood frames. Put them in hive *B*, in which you have previously placed: 1) at one end of the hive, a honey frame; 2) next, the two brood frames; 3) five or six frames, built out with comb but empty. Close hive *B* and put it in the former location of hive *A*, which should be moved a good distance

---

* Called a "walk away split" in the U.S. today. *Ed.*

CHAPTER 17. ALTERNATIVE PROCEDURES

away (fig. 200). The queen will now be left in one hive or another; the one without a queen will appear agitated at first, but will build some new queen cells.

There are other methods of artificial swarming that require finding the queen, but these methods are more difficult, and in any case are not the best.

235. Other procedures for uniting colonies

As we've seen, managing fixed-comb hives often requires uniting colonies or swarms, while managing moveable-frame hives usually allows you to avoid such procedures.

However, we have provided one method for uniting colonies for moveable-frame hives (§ 132), and we've shown how to unite fixed-comb hives (§ 204). Here are some other methods that can be used for all hive systems.

1. *Uniting using ether.*[*] Soak two pieces of cotton, roughly 3/8" x 3/4" (1 cm x 2 cm) in size, with ether, then place one of the pieces beneath each of the hives you plan to unite. Leave them there for twenty minutes or so, then unite the hives as usual—now without the danger of fighting among the bees.

2. *Uniting using flour.* After smoking the bees, throw a few handfuls of flour in between the combs of the two hives you plan to unite, then unite them—without fighting.[**]

236. Restoring queenless hives

As we've seen, the simplest thing to do when a queenless hive is discovered is to get rid of it. This is always true of orphaned hives found during the spring or fall inspection.

---

[*] This system is recommended, and has been used successfully, by Mr. Bourgeois, who lives near Chartres. (Ether is not commonly available in the U.S. today due to potential for drug abuse. *Ed.*)

[**] Some beekeepers use powdered sugar to the same effect. *Ed.*

But if you find that a strong hive has become queenless during foraging season, you can try to give it a queen using one of these methods:

1. *Adding brood of all ages.* As we know, a colony that has recently lost its queen can produce a new one using extremely young larvae.

If you're not too far into foraging season, you can add comb containing brood of all ages to the queenless hive—along with a comb of capped brood to bolster its population. When you inspect it, eight to ten days later, you can check to see whether the colony has built any queen cells.

If it has managed to produce a new queen, then you will have restored a colony capable of surviving the winter, and even prospering the following year. If there is no new queen, you can try once again by adding another comb with brood of all ages—and if the bees fail once again to build queen cells, you can unite this colony with another.

2. *Adding a reserve queen.* You can try to have a queenless hive accept a new queen, going about it much as you would when requeening (replacing queens artificially, § 237). This delicate and difficult procedure only has a chance of succeeding if the hive has only been orphaned for a short time—and even then, it often fails.

237. Artificial requeening

Instead of allowing queens to be replaced naturally ("supersedure," § 158), it is possible to replace an existing queen with another (that is, to "requeen").

As we've just seen, you can also give a new queen to an orphaned hive, or to a colony from which you've taken an artificial swarm.

In any of these cases, you'll need, in one way or another, to have a certain number of queens at your disposal.

The following presents two methods you can use when you don't wish to allow queens to be replaced naturally, or to dissolve orphaned hives during foraging season.

Finding the queen. Sometimes these procedures require locating the queen inside a hive.

CHAPTER 17. ALTERNATIVE PROCEDURES

To do so in a moveable-frame hive, you'll need to examine it frame by frame, beginning with those containing brood. Before you begin, smoke *lightly*.

Only light smoking is required in this instance, because if you smoke too much, until the bees produce a loud humming, the queen may leave the combs altogether and hide in a corner of the hive, where she'll be very difficult to find.

Another, easier way to locate the queen in a moveable-frame hive is to place an empty frame built out with worker cells into the hive the day before you plan on looking for the queen; this empty frame should be placed in the middle of the brood frames.*

The following day, the queen is likely to be found laying eggs on this frame, and, with a bit of luck, you'll find her right away.**

---

* If you want to search for the queen directly, you'll need to examine the brood combs one by one. But sometimes a queen who has been frightened by the procedure will hide in a corner of the hive. In this case, if you're unable to find her, spread a cloth on the ground, and place an empty fixed-comb hive on top. Shake out all of the frames with bees on them, one by one—and, as the bees make their way toward the empty hive, look for the queen. As soon as you see her, trap her beneath a glass, turned upside-down.

** Here's a method that may seem drawn-out, but which often proves quicker than looking for the queen directly. Place an empty moveable-frame hive (without a bottom board) atop a sheet spread on a nice, flat spot on the ground. Replace the metal tab at the entrance with some perforated sheet metal that will allow worker bees to pass through, but not the queen (fig. 201). Remove the brood frames one by one from the hive whose queen you're looking for—and do so almost without smoking at all in order to avoid spooking the queen. Shake the bees out onto the ground in front of the entrance of the moveable-frame hive, inside of which you should place the frames shaken free of bees. Using a bit of smoke, drive the workers through the perforated metal, which they'll cross in an attempt to make it back to their comb. The queen, however, will be stopped by the perforated metal, and will be easy to locate.

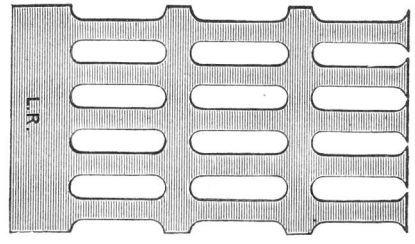

Figure 201. Piece of queen excluder—perforated sheet metal that allows worker bees to pass, but not the queen (1/2 actual size).

238. Requeening by natural swarming

You can maintain a certain number of low-capacity fixed-comb hives as a kind of *breeding apiary* or a *nursery*, to produce new colonies for reinforcing your apiary, and to furnish new queens through afterswarms.

If, instead of returning an afterswarm to its hive, you collect it and turn it into a new colony, the resulting colony will have a new queen who will typically be highly fertile.

When a second, or even a third swarm, is installed in a moveable-frame hive with just a few frames, you'll have this small colony's queen at your disposal.

If you have several apiaries, this method is perhaps the best for maintaining them, since your nursery apiary will always contain small colonies that you can always add to others that, for whatever reason, are disorganized to some extent, or queenless.

However, one inconvenience of this method is the need to watch for and collect the swarms that occur in the nursery apiary.

239. Requeening by grafting queen cells. Making nucleus hives

In early spring, choose one of the best colonies in your apiary, and stimulate egg-laying by uncapping a few honey combs. When the colony has become very strong—that is, when it contains between eight and ten brood frames—locate its queen. As we've just seen, this can be a difficult task. When you find her, remove the frame she's on, along with any other bees on it. Place this frame in a hive along with a honey frame and eight to ten frames built out with comb. This hive should then be placed in a spot previously occupied by another strong hive, which will be moved elsewhere.

The hive whose queen you've removed will naturally build some queen cells.

Around seven days after removing the queen, count all of the fully formed queen cells (two cells that are joined together only count as one). Then, prepare some small *nucleus hives*, equal in number to the number of queen cells, minus one.

To create these nucleus hives, you'll need to have some ordinary

CHAPTER 17. ALTERNATIVE PROCEDURES

hives on hand, or, preferably, some small boxes that can only contain three frames.

From every strong colony in your apiary, remove one frame of brood, along with the bees on it—but without the queen. Place each of these frames in one of your nucleus hives, in between two frames, one of which should contain honey and pollen, then sweep the bees from a third frame from the same hive (but not the queen!) in front of the nucleus hive entrance as described in § 217 with the observation hive. Store these nucleus hives in a cellar for 48 hours, being sure to screen the entrance. Then, place them in their final locations, opening each entrance just wide enough for two bees to pass at a time.

On the tenth and eleventh day after removing the queen from the first colony, open that hive and cut out a triangle around each queen cell, except for one, which should be left in place to allow the original hive to replace its queen. Cut very carefully to avoid damaging the queen cells.

Place these excised queen cells one by one into a box lined with cotton, protected from sunlight and cold. Handle them with great care, and avoid shaking. Then, cut out a similarly shaped triangle from the middle of the brood comb in each nucleus hive, and carefully put in its place one of the queen cells, with the same orientation as it had naturally in the parent hive (fig. 202).

Later, check to see that the nucleus hives contain brood. If they do, then you now have at your disposal a young and fertile queen that you can give to your various colonies as necessary.

Any nucleus hives that contain no brood should be added to other hives.

Sometimes these small nucleus colonies may abandon the hive together with their queen when she first leaves the hive. To avoid this, you can take a second brood frame, along with the bees on it (but without the queen), from another colony and add it to the nucleus colony.

If you've maintained several of these nucleus hives until the end of the season without making use of their queens, or if, despite their growth, they've been unable to form sufficient winter reserves, you'll need to combine them with one another, or with other hives, after eliminating the older queen.

Figure 202. A beekeeper grafting a queen cell onto a frame.

As you can see, this method is extremely complicated and brings with it a number of disadvantages—not to mention the risk of robbing, which is a real danger during the various procedures involved, along with the risk of foulbrood originating in the small hives, where, despite all precautions, there are sometimes not enough bees to cover all of the brood.

240. Introducing a queen into a hive using a queen cage

Let's assume that a hive has a deficient queen, which you've located and eliminated.

Take a small tube made of metal cloth, closed on the top and bottom with corks (fig. 203), and insert a new queen taken from a nucleus hive, holding her by the wings. Move apart the frames in the hive whose queen you've eliminated, in order to wedge the metal tube containing the young queen in between two brood frames. Be sure that the cage is positioned within reach of a section of honey comb, uncapping any capped cells that may be there, so that the queen can feed on the honey by extending her proboscis through the metal cloth; then, cover the frames up again as usual.

Two days later, uncover the frames and gently remove the cage, along with any bees that are on it; smoke lightly, to avoid frightening the small cluster of bees.

If you notice workers trying to sting the queen or pulling her legs, put the cage back where it was, and repeat the inspection several days later. If you see workers trying to feed the queen with their proboscises, drive away the workers on the cage and replace the bottom cork with a small piece of comb, then put the cage back where it was. The workers will chew through this bit of comb and free the queen themselves, in which case she'll usually find acceptance.

Once the queen has been freed, remove the cage and put the frames back exactly as they were. A few days later, check to see whether the queen has been laying eggs.

There are multiple varieties of queen cages (figs. 204 and 205); the simplest is the one we've described here.

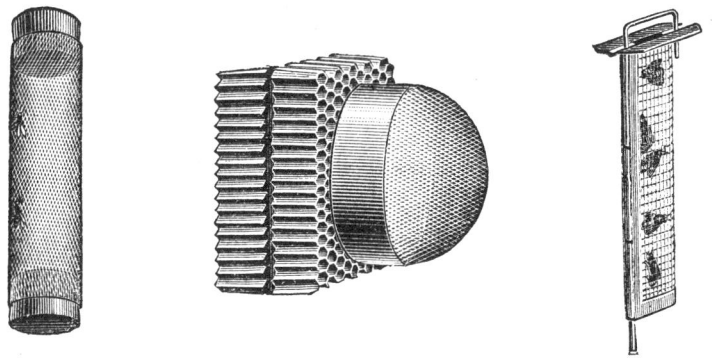

Various queen cages.

Figure 203. A tube cage.   Figure 204. A bell cage.   Figure 205. A flat cage.

241. Introducing a queen into a hive through the entrance

In some cases, despite all possible precautions, a queen may not be accepted, and be killed by the bees.

Here's another, extremely simple procedure that seems highly effective* and has the advantage of being equally applicable to fixed-comb hives and moveable-frame hives.

Figure 206. An atomizer.

Take a hive whose queen you've removed and now must replace, or a hive that has been orphaned, smoke it heavily, and sprinkle the combs with sugar water containing a few drops of essential oil.

Next, take a frame or two from the hive and brush off the bees in front of the entrance. Then, take the queen out from under a glass—where you placed her temporarily—and throw her right in the middle of this crowd of bees, who will pass their perfumed odor along to her as well. An atomizer may prove useful here (fig. 206). The bees will reenter the hive along with the queen, whom they will now easily accept. This procedure should be carried out around nightfall to avoid robbing.

242. Foreign bee races

There are a number of species, sub-species and varieties of bees, each with its own positive and negative traits.

Among these many varieties, *Italian* and *Carniolan* bees have been adapted for use in a range of countries and kept just like ordinary local bees.

Attempts have also been made to keep, in the same way, bees from Palestine, Syria, Cyprus, and Egypt—but since these four varieties are extremely difficult to handle, and often attack their keepers, they are rarely used anymore.

Purebred Italian and Carniolan bees don't have this disadvantage, but Carniolan bees are highly predisposed to swarming, even when housed in extremely large hives.

---

* This procedure is recommended by Froissard.

Experts often recommend keeping Italian bees due to their high activity levels, and to their slightly longer proboscis, which allows them to visit certain flowers that other bees can't.

However, these advantages fall far short of compensating for the following drawbacks:

1. It is all but impossible to maintain the purity of the Italian race in an apiary, since Italian bees crossbreed easily with ordinary bees. Even if you eliminate all indigenous black bees from your apiary, crossbreeding can easily occur with bees from the surrounding area, even from several miles away.

    And the hybrid bees that result from such crossbreeding are often aggressive and vicious.

2. Italians are particularly prone to robbing, and require a good deal more supervision in this regard.
3. An Italian queen bought remotely, and produced in unknown conditions, may infect the hive with foulbrood.

So adding foreign bees to your apiary can hardly be recommended.

Quite the contrary: beginning beekeepers should be warned against falling for passing fads that can sometimes cause serious harm.

It must be said that some foreign bees—notably, Italian bees—look great, and are wonderful to watch at work. If this pleasant spectacle appeals to you, it may certainly grace your apiary—it's up to you!

Introducing a queen of a foreign race. A queen of a foreign race* can be introduced using the two methods already described (§§ 240 and 241). The first method can be modified as follows for better results.

The best way to introduce a queen is to first prepare a queenless colony that has only young bees. A few days before introducing the foreign queen, take three frames full of bees (but without the queen) from a strong colony—they should contain honey, and one of them should include brood. Place these frames on one end of an empty hive; the frame with brood should be placed in between the two others.

---

* Foreign queens are shipped in small boxes, which usually contain a piece of comb and several worker bees.

Next, move the hive into a cellar or a dark room. After two or three days of isolation, move it to any location in the apiary and open the hive entrance.

Here's why these steps are useful: experience has shown that a small colony composed primarily of young bees will more readily accept a foreign queen than a colony with many older bees. So we first created a small colony, and once it is returned to the apiary, it will lose most of its older bees, who, out of habit, will return to their previous hive. Isolating the hive in the cellar helps the bees become accustomed to their new home.

Now, once you have a new queen ready to introduce, open the small colony. After destroying any queen cells currently under construction, and putting those frames back into place, slide the queen cage containing the foreign queen in between two frames. The cage can be kept in place easily—just wedge it in between the two combs. Now, proceed as described previously (§ 240).

243. Comb honey without sections

Regardless of which system of hive you're using, when you have some nice combs, built without foundation and filled with capped honey, you can prepare comb honey as follows.

Using a shaped comb-cutter—much like a cookie-cutter—cut out some pieces of this honey comb, with the desired dimensions, and place these attractive bits of comb in tinplate boxes specially made for this purpose.

In this way, you can sell such highly attractive comb honey, while avoiding all of the inconveniences of producing section honey.

244. Partial harvest from a fixed-comb hive

Many keepers who maintain fixed-comb hives and don't harvest using the suffocation technique still use a different method than the one we described (§ 201). They use a technique called *partial harvest*.

Here's how they do it.

Smoke the hive from below, then turn it upside-down and move it some distance away, where all of the necessary tools are at hand. Take

CHAPTER 17. ALTERNATIVE PROCEDURES

an appropriately sized board and cover the combs occupied by brood and half of the adjacent combs on each side of the brood nest. Use some smoke to drive the bees underneath this board, thus leaving the honey combs uncovered. Cut out these combs, remove them with a knife (fig. 207), and sweep off any bees that happen to be on them. Then, after shaking any remaining bees off the board into the hive, put the hive back in its place.

It's best to carry out this procedure while there is still a honey flow—otherwise, robbing may break out.

Figure 207. A beekeeper cutting comb from fixed-comb hives.

Summary

Alternative procedures
There are some beekeeping procedures that can be used as alternatives or as supplements to those we described earlier.

Buying hives. Transfers. Feeding
Buying hives: if a beginning beekeeper manages to acquire pre-populated moveable-frame hives, this will certainly make it easier to get started.

Transfers: if you have small fixed-comb hives, you can transfer them by placing them on top of moveable-frame hives.

Feeding: 1) what is usually called *speculative feeding*, meant to produce a strong population in advance of an anticipated honey flow, usually brings more problems than advantages, and can only be recommended for use in regions where the weather can be foreseen with a high degree of certainty; 2) if you have to feed very late in the season or even during the winter, it's a good idea to do so using a paste made by mixing honey and powdered sugar.

Swarming. Uniting hives. Queenless hives

Preventing afterswarms: you can relocate the hive when you hear the queen's song. You can also remove all of the queen cells except for one, although in this case you risk orphaning the hive.

Artificial swarming: a beekeeper can produce artificial swarming using just one moveable-frame hive.

Uniting colonies: whether you're using fixed-comb hives or moveable-frame hives, you can combine them without fighting among the bees by using cotton soaked in ether or by sprinkling the bees with some flour.

If strong hives have become queenless during the course of the season, you can try to restore them, either by adding brood of all ages or—which is more difficult—by introducing a reserve queen.

Requeening. Foreign bee races

Generally, beekeepers let their bees replace their queens naturally. If, however, you want to replace queens you know are deficient yourself, you can do it as follows: keep a few fixed-comb hives as a kind of nursery apiary, which serves to support your actual apiary using natural swarms. Collect the afterswarms, which will furnish virgin queens. There's also another way of replacing queens—by grafting queen cells in small queenless hives established for this purpose. However, this procedure is extremely complicated.

When it comes to foreign bee varieties that keepers have tried to keep like ordinary local bees, only the Italian bee is reasonably common. You can introduce an Italian queen (as you'd replace any other queen) by eliminating a hive's existing queen; then, using a queen cage containing the new queen, or by mixing her in with a crowd of bees you've swept in front of the hive and sprinkled with perfumed sugar water. Using Italian bees can't be recommended, since their disadvantages outweigh their benefits.

Partial harvest of fixed-comb hives

When it comes to harvesting fixed-comb hives, a partial harvest technique is sometimes used, which consists simply of turning a fixed-comb hive upside down in order to cut out the honey combs.

PART FOUR
GENERAL OBSERVATIONS ON BEEKEEPING

Chapter 18
General Principles And Comparison of Methods

245. Preliminary remarks

Before comparing the various procedures described in the previous chapters, it's a good idea to tease out a few general principles that are of primary importance for any beekeeper.

First, we should draw a clear distinction between *hive systems* and *beekeeping methods.*

You can be a good beekeeper with any system of hive, but you can't be a good one with just any method.

As for equipment, we've seen that, based on circumstances and the goals you've set for yourself, the three most widely recognized types of bee housing are: traditional fixed-comb hives, horizontal moveable-frame hives, and vertical moveable-frame hives. We've described the best models in all three categories.

246. General principles applicable to all hive systems

Regardless of which hive system you choose to adopt, you should always strive to follow these general principles as closely as possible.

I. Make sure that your region will provide sufficient honey flow before going to the trouble of establishing a sizeable apiary

This may seem obvious—but we simply can't stress it enough. Many beginners either are unable to assess their region's honey flow, or even set up an apiary without giving any thought to this question at all. Some beekeepers may manage a fairly large number of hives in a region with mediocre honey flow, and see no honey production whatsoever. Surprised by this futility, beginners may even blame the methods they've been advised to use, or the hive system that was recommended to them, when in fact the lack of honey production is due, quite simply, to the scant honey flow in the surrounding area.

If there's no nectar in the surrounding plants, there will be no honey in the hives—it's that simple.

II. To harvest as much honey as possible, you need high-capacity hives

If you're keeping bees in order to harvest honey, and not for breeding purposes, you need to be in an area with sufficient honey flow—and in this case it's a good idea to install your bee colonies in large hives. The volumes we've recommended for the three types of hives (horizontal hives, vertical hives, and cap hives) are usually the best. Also note that the size of the hives should be expanded further in proportion to the greater honey flow of the surrounding region.

The major advantages of large hives are as follows:

1. Bees housed in a large hive will swarm less frequently than bees in a small hive,* and, as we know, managing an apiary is made easier by preventing, or at least reducing, natural swarming.
2. With more room available for brood development, you're more likely to see a larger population by the start of honey flow.
3. The completed combs, still empty of honey, take up a larger surface area when the honey flow begins, allowing the bees to spread the honey they've gathered across a greater surface in order to promote evaporation of water.

III. To harvest as much honey as possible, you need strong colonies

Simply having a large hive isn't enough to guarantee that the colony housed there will develop a large population. This depends first and foremost on the queen's fertility. As we've seen, regardless of what hive

---

* Unless you're dealing with certain bee races that can never be prevented from swarming.

system you've adopted, the procedures a beekeeper performs should always aim to strengthen colonies.

This is all based on the fact that the labor performed by a colony whose bees weigh 12 lb (6 kg), for example, is much greater than twice that performed by a colony weighing only 6 lb (3 kg).

IV. During the normal course of managing an apiary, you should avoid working the bees too frequently

As we've seen, a beginning beekeeper is advised to inspect the bees frequently as part of his learning experience—but the opposite advice is appropriate for an established beekeeper. Any method that manages to involve as few procedures in the apiary as possible, without reducing the honey harvest, is, for this very reason, a good one. Why? Because the two dangers a beekeeper fears the most—robbing and foulbrood—are more likely to be prevented the less you handle your bees.

For this reason, it's best to avoid having to unite colonies or feed hives, and to allow queens to be replaced naturally.

V. You should always assume the worst when looking ahead to the next season

If you always keep this idea in mind, and successfully resist the temptation to harvest too much honey, you'll be sure to avoid the kinds of setbacks that could eventually make you give up beekeeping altogether.

To brace yourself for a bad year—which you should always assume is coming—and to avoid all of the trouble of feeding, you should always have comb with capped honey at the ready.

If you're using moveable-frame hives, then frames with comb full of capped honey will make up this reserve. If you're using cap fixed-comb hives, you'll need a number of caps that you've set aside for a rainy day—but for this kind of hive, such a reserve can only be used for springtime feeding.

247. To what extent should bees be allowed to build comb?

One of the major advantages of moveable-frame hives is that they allow the beekeeper to have on hand a large number of frames, fully built out with comb, ready for the bees too fill rapidly with honey. Using an extractor, you can remove the honey without destroying the comb, which the bees can once again fill with honey.

Does it follow that bees should never be allowed to build more new comb, and that the beekeeper should work to suppress, once and for all, their natural inclination to build?

Numerous and rigorous experiments have proven that, during the spring and the primary honey flow, it's to your advantage to leave a certain number of primed frames for the bees to build out as they see fit, along with the large number of already finished frames you give them.

All else being equal, a hive that is arranged in this fashion will yield just as much honey by season's end as one whose comb was already completely built at the start; and, on top of the honey, you'll have the wax that the bees have produced.

This is the system described in § 161.

248. Protecting a colony against fluctuations in temperature

As we've seen, during wintering a lack of ventilation is a greater cause for concern than the cold. But that's not to say that steps shouldn't be taken to protect bees against cold during the winter—especially against drastic changes in temperature.

Such fluctuations are even more harmful during the spring, when the brood is already developing in the hives: a sudden cold snap forces the bees to cluster, and thus abandon some of the brood, which in turn could develop foulbrood.

Some beekeepers have had the idea of building hives with double walls. Such hives are excellent, but they're rather costly, and experiments have shown that simply covering a hive with a straw mat gives the same level of protection as an extra wall.*

---

* For more details, see G. de Layens, *Nouvelles expériences pratiques d'apiculture*, p: 19 (Paul Dupont, editor).

Another idea for preventing heat loss in a moveable-frame hive is inserting division boards; as we've discussed (§ 227), experiments have shown that wax comb, fully built-out but empty, insulates just as effectively as a partition board, which can therefore be considered a useless complication.

All of this means also that it's harmful to disturb your bees frequently during the spring by adding frames in several installments, as was advised once upon a time.* It's much better to fill a hive completely with built-out or primed frames. This way, even without a division board, you achieve much better insulation, since each extra built-out frame you add will do the work of a division board.

249. Various kinds of beekeepers

Depending on their situation and their aims, beekeepers can be grouped in several categories. This being the case, the methods and equipment recommended for each category won't always be the same.

We can distinguish between the following:
1. Beekeepers whose jobs take up most of their time, making beekeeping a side interest. This would include farmers who are concerned above all with their crops, or beekeepers whose professions demand most of their attention—businessmen, doctors, pharmacists, civil servants, etc. We might call them *sideline beekeepers*.
2. Beekeepers who wish to be productive enough to make beekeeping their primary occupation. We'll call them *professional beekeepers*.
3. Beekeepers who have a lot of time to dedicate to bees, and who, so to speak, are more interested in the bees themselves than in their potential products. We'll call them *amateur beekeepers*.

---

* Many beekeepers find that adding frames to a horizontal hive, even in multiple installments, barely disturbs the bees at all. This is a great advantage of the horizontal hive with frames' top bars forming a ceiling. Beekeepers in northern climates often find it beneficial to limit the hive volume with a division board until reliably warm weather sets in. *Ed.*

250. The sideline beekeeper

Anyone who has only limited time to spend on bees will have to choose a method that requires less monitoring of the hives, while still providing as large a honey harvest as possible.

The horizontal moveable-frame hive is best suited to this purpose, using the method summed up after § 170.

But establishing an apiary of this kind can require an investment that not every aspiring beekeeper can muster. In this case, you can continue using the fixed-comb hives you used during your apprenticeship; later, you can begin adding moveable-frame hives, purchasing them with money earned by the honey harvest during good years.

Why not simply recommend that a beekeeper in this situation continue using fixed-comb hives indefinitely? Because, truth be told, managing fixed-comb hives well requires more time and apicultural experience than doing so with horizontal hives.

Why not recommend vertical moveable-frame hives for sideline beekeepers? Because vertical hives require more monitoring and more delicate procedures than horizontal hives, without yielding better harvests.

251. The professional beekeeper

Professional beekeepers will aim to do one of three things, depending primarily on the kind of region they're in: produce extracted honey, produce section honey, or pursue breeding and produce bees for sale.

1. If your aim is to produce extracted honey, you should adopt moveable-frame hives, since they'll allow you to take more or less full advantage of periods of intense honey flow, and to harvest, with an extractor, honey that is purer than that produced with fixed-comb hives. If time is not an issue, you can take your pick between horizontal and vertical hives.

However, if as a professional beekeeper you require a large number of hives, you should avoid placing all of them at a single location. Quite obviously, there's a certain maximum number of hives you can set up at any given spot, since the excessive population of bees will be unable to gather enough honey for each individual hive. Experience has shown that, for a region with reasonably abundant nectar sources, it's wise to stay below the fifty colony mark—and the apiary shouldn't be near any other sizeable apiary. So a professional beekeeper should distance his hives several miles from one another.*

2. If professional beekeepers happen to be in a region where producing section honey can be profitable, they should use vertical hives with shallow frames.

3. If a professional beekeeper is engaged in breeding, it's preferable to use fixed-comb hives, since this is the form in which populated hives are typically sold.

If you're certain to be able to find buyers for populated moveable-frame hives, then you can pursue breeding using these hives as well.

252. The amateur beekeeper

An amateur beekeeper whose main objective isn't his apiary's output can, of course, adopt any hive system without any major concerns, and can make use of all supplementary beekeeping equipment that many more practically-minded beekeepers could easily do without.

The best advice for an amateur beekeeper, in terms of the best interests of beekeeping generally, is to dedicate a part of your time, once you've mastered the trade, to setting up careful and thorough experiments to investigate various issues in beekeeping for which a clear solution has yet to be discovered.

---

* See chapter 24.

Summary

Depending on whether you're a sideline beekeeper, a professional beekeeper, or an amateur beekeeper, there are various methods and hive systems you can adopt—but all beekeepers, regardless of their particular aims, should keep the following principles in mind.

General Beekeeping Principles

1. Be sure that your region has sufficient nectar resources before setting up hives.

2. Keep large-capacity hives.

3. Maintain strong colonies.

4. With the exception of your period of apprenticeship, avoid handling your bees often.

5. Always manage your apiary as if the next season will be extremely poor.

6. To manage your bees in as orderly a way as possible, avoid natural swarming as much as possible.

7. Take every step possible to ensure that your hives are ready for wintering, leaving your bees a honey reserve large enough to last the entire cold season, and ventilating each hive properly.

If you stick to these principles, you will avoid—as much as possible—the inconveniences of feeding, robbing, and foulbrood. And, if you form a *capped honey reserve*, you'll have something to fall back on during those bad years that could otherwise cause irreparable damage for a beekeeper with less foresight.

Figure 208. A beekeeper's honey room.

Chapter 19
Apiary Products

253. General considerations

The main product of an apiary is, of course, *honey*, which can be sold in the form of *extracted honey*, or—less frequently in France—*comb (or section) honey*.

Despite its high value, *wax* is only a secondary product, since, as we've seen, it's in beekeepers' best interest to keep a large stockpile of built-out comb to add to hives later for the bees to refill.

Finally, in certain regions, a beekeeper may profitably sell bee colonies—that is, engage in *breeding*.

You'll need to give some thought to maximizing your profit from the various products your apiary can generate. Thus, for example, some beekeepers simply sell a number of fixed-comb hives—thus selling the honey, the bees, and the wax all at once, without having to worry about extracting the honey or melting down the wax. But, generally speaking, this isn't the best way of doing things.

Extracted honey is the easiest product to sell in France, but in certain areas where honey production is on the rise, selling it can sometimes become difficult. In this case, it may be beneficial to use a certain portion of your honey to make *mead* (§ 258), either for your own consumption or for sale.

As for comb or section honey (§ 191), we've seen that, in France, one must be certain ahead of time of finding a market before going about producing it profitably.

We should mention that there are many secondary uses of honey (§ 276): you can use it to make mead, brandy (§ 275), and vinegar (§ 274) that you can be certain is pure in origin.

254. The honey room

The simplest kind of honey room a beekeeper can have is any room whose doors and windows can be tightly sealed to prevent bees from entering. If the room has a chimney, you should even be sure to seal it with some wire mesh to prevent the bees from getting through. In any event, the room, equipped with tables or boards, should be large enough to accommodate all of the necessary procedures. It's best to have a dedicated space—not to use a room that also serves other purposes.

The room you choose won't just be a place to carry out the procedures we've discussed, but will also store your harvested honey and frames, including those containing reserve honey, all of your as yet unused comb, wax to be melted down, your extractor, and all of your other beekeeping equipment.

Since this room will hold honey and frames in various states, it's absolutely essential to maintain a constant air current in the room to prevent mold from developing. The simplest way to establish this current of air is to drill two openings in the walls, opposite each other, each screened with some wire mesh to keep the bees out.

In this honey room, install shelves for holding frames with empty comb or full of honey, to be placed vertically on these shelves (fig. 208). The tops of the frames are kept at a small distance from one another by nails driven into the bottom of the shelf above.

The honey room will also allow you to sulfurize all of your frames at once: just carefully close all of the openings, and burn some sulfur in a tray in the middle of the room.

If you so desire, you may want to set up a more elaborate honey room. Take, for example, one highly convenient and simple addition that can make it easier to drive away bees when they congregate around the windows: installing a window that swivels around a central axis, whether horizontally or vertically. When a certain number of bees have accumulated on the inside of the windowpane, you can swivel the entire window one hundred and eighty degrees, and all of the bees will find themselves outside.

255. Storing honey

Honey abhors humidity, and should be kept in a dry and well-ventilated area—ideally in airtight jars.

One very important point that should never be forgotten is that honey should always be harvested from comb that is for the most part capped. If you harvest honey from uncapped comb, then the resulting honey—which contains much more water than mature honey—will be difficult to store, and prone to fermentation.

After a certain time, most varieties of honey begin to crystallize. This crystallization process begins with the formation of grains that soon affect the honey's color. The granulation increases gradually, with some liquid still interspersed between the grains; the honey takes on a paste-like consistency. Then, after a certain time, it will typically become quite hard.

If honey that has remained liquid on top draws enough water from the humid air, then this upper portion can ferment. You can drain it off, then hermetically seal the rest of the honey.

Some varieties of honey, even when capped only recently, have a consistency that causes them to stick tightly to the cells. These include, for example, heather honey, which cannot be removed using an extractor. In this case, you have no choice but to destroy the comb, or to remove the honey as previously described (§ 167).

Different varieties of honey are more or less prone to crystallization. Some honeys, like rape honey, crystallize rapidly; others, on the contrary—like that of pure sainfoin—crystallize more slowly. Generally, honey derived from a variety of flowers is most likely to crystallize well.

256. Selling honey

Honey will sell more easily when the buyer can be certain of its origin—and therefore of its purity. The question of "confidence" plays an important role in selling this product.

Needless to say, the honey should be presented attractively in any case, with labels indicating the apiary it came from.

Because of this concern with origin, the sale of honey, more than that of any other product, should exclude middlemen as much as possible.

A consumer or merchant will often pay more for a mediocre honey with a known origin than for a superior honey offered by an unfamiliar seller.

It's difficult to give relative price points for various kinds of honey, because consumers living in a certain region are often accustomed to the taste of the local honey and may prefer it to any other.

257.1 Major varieties of honey

A honey collected in mountainous regions—a white honey known in the Alps as *Chamonix honey*—is the most highly valued and, typically, the most expensive honey.

Sainfoin honey is one of the most highly valued in the northern and central regions of France, for example, where it's a fine-grained white honey known as *Gâtinais honey*.

The honey of the Mediterranean region, from Provence and Languedoc, for example, is highly aromatic—*too* aromatic to Northerners' taste—but it's preferred by Southerners, who are less fond of sainfoin honey. Mediterranean honey can vary in color. One of the best and most refined honeys is known as *Narbonne honey*.

Heather honey is reddish-brown in color, highly viscous, and with a taste that is not widely appreciated; it's less valuable than the previous honeys. This honey is known as *Landes honey*.

Buckwheat honey is similar in color, but less viscous; it is also lower in quality and has a rather unpleasant taste. An example is the so-called *Brittany honey*.

The latter varieties of honey, which are often blended and referred to as *red honey* or *dark amber honey* (*miel rouge*), are used primarily to make gingerbread, which makes them quite easy to sell.

There are many other kinds of honey that we won't mention here.

257.2 Honey composition

Not all honeys have the same composition, which depends on the makeup of the various sugary substances collected by the bees. Later (in § 297), you'll find some notes on how to analyze nectar and honey flow, as well as how nectar is transformed into honey.

Generally, capped honey contains 25% water,* a large percentage of glucose and fructose, a smaller percentage of sucrose (cane sugar), and a small amount of dextrin.

Here, for example, is an analysis of honey derived almost entirely from sainfoin, just after uncapping.

SAINFOIN HONEY

Water	22.54
Sucrose (cane sugar)	6.10
Glucose and fructose	69.26
Dextrin	0.07
Gums, mineral materials, and loss	2.03
	100.00

Honey gathered in the mountains contains a higher percentage of sucrose—it can sometimes exceed 10% of capped honey by weight.

Meanwhile, some honeys, like heather honey, contain almost nothing but glucose and fructose; some honeys have a significant percentage of dextrin; and capped honey derived from honeydew may contain as much as 5% dextrin by weight.

---

* Generally, capped honey should be no more than 18.6% water. Above 19 to 21% it may ferment. Excessive moisture can be driven out of harvested frames by ventilating uncapped combs for a few days in a dry space at room temperature. *Ed.*

Below are a few analyses of honey provided by Gayon,* a professor at the Faculté des Sciences of Bordeaux.

Origin	Sugar content, %		
	Sucrose	Glucose and fructose	Dextrin
Eure	8.00	66.60	0.10
Lot-et-Garonne	5.02	71.00	0.06
Vendée	2.14	73.50	1.03
Gironde	12.92	61.00	0.20
Aisne		78.10	7.29
Switzerland	5.60	67.60	4.32
The United States	7.69	71.40	0.45

258. Mead

Mead or *honey wine* is an alcoholic beverage made by fermenting honey added to a certain amount of water; it's a drink popular among Slavic nations. Excellent mead is produced in Russia, Dalmatia, and Poland, for example.

The interest in producing mead stems largely from the fact that honey can be difficult to sell in regions where it is produced in great quantities. In this case, it's in a beekeeper's interest to produce—at the very least for his own consumption—this excellent drink, which can rival the finest white wines or Spanish wines in quality. With honey at an average price, mead with 13 to 17% alcohol content can cost between 0.30 and 0.50 francs per liter. This is, of course, a very lucrative price, particularly since many inferior varieties of honey can produce perfectly respectable meads.

The question of whether mead can be produced for sale is only now beginning to be widely discussed; surely answering this question in the affirmative will give a major push to the development of beekeeping in the future.

---

* See *Apiculteur*, 1892, p. 298.

259. Low-quality mead

In France, particularly in the North, there is a sweet liqueur that is known as mead, but exhibits none of the qualities worth appreciating in fine honey wine.

Truth be told, a great many formulas for mead production have been proposed, and one often hears a particular formula promoted for no apparent reason. This confusion helps explain why many beekeepers fail to produce good mead. In § 262, you'll find a very simple, fail-safe procedure for producing a top-quality mead, provided that the procedure is followed exactly.

260. Alcohol content of a fine mead

The first attribute of a good mead is a high alcohol content, 15–17%. And strong meads have a further advantage: you can store them indefinitely. As they age, they can rival the finest wines in quality—and such aged meads are those most appreciated by tasters.

Meads with lower alcohol content lack these qualities.

But a high-alcohol mead can be diluted perfectly well with water—much better than any white wine. This means you can serve a strong mead as an everyday table beverage by cutting it with a sufficient quantity of water.

The taste of a good, strong mead, sufficiently aged, no longer bears any resemblance to that of honey.

261. The bouquet and color of mead

Just as the quality of various honeys varies based on the aroma of the flowers they are made from, the taste of mead varies depending on the honeys used to produce it. You could even say that meads come in various vintages, just like wines do.

Mead produced using colored honey loses its honey-like taste less rapidly than mead made with white honey; but, as it ages, it is often superior in terms of bouquet.

For example, a mead containing a certain amount of heather honey can become remarkably good with age, and is as good as some

Spanish wines. However, a beekeeper can only rely on experience when it comes to selecting honeys.

White wine with no coloration at all is usually not highly regarded—so you'll need to appeal to both the eye and the palate of the consumer. If the mead is transparent, you'll want to color it using caramel syrup. Such syrup is widely available commercially, and is used to color low-grade brandies. All it takes to lend the liquid a nice golden tint is to add some—around a small Bordeaux wine glass full (0.5 L) per 100 L—to a barrel during fermentation.

262. General production method

Take, for example, a 100-L barrel. Pour in 25 L of honey—the equivalent of approximately 37 kg*—then pour in 74 L of water.

The barrel should not be filled precisely to the top, since otherwise the liquid would spill over during the first fermentation; so leave approximately 1 L of empty space. Next, put 50 g of tartaric acid into the barrel (this serves to activate fermentation), along with *10 g of bismuth subnitrate, which serves to prevent secondary fermentation—a very important point.* These products are available in any pharmacy.

Finally, remove comb from a hive, containing that year's pollen, and put around 50 g of it into the barrel, after first thinning out the pollen with a bit of the liquid from the barrel;** the pollen serves to nourish the fermentation process with a nitrogenous substance. With a stick, stir up the liquid, mixing everything thoroughly.

The three products just mentioned are essential if this procedure is to succeed.

The only thing remaining to do is to place some water-soaked fabric atop the plughole, and, on top of that, some wet, firmly packed sand.

You can tell that the fermentation process has died down if you no longer hear a fizzing sound when you place your ear against the barrel. When this happens, replace the fabric and sand with the plug.

---

* If you're using crystallized honey, then place the jars containing it in front of the fire to melt it first.

** These proportions are approximate; it's no big problem if the bits of wax are mixed up with the pollen you add.

From this point on, you needn't do anything to the mead until you're ready to bottle it (see § 266).

If you taste the mead during the production process, particularly near the end of fermentation, you'll often find it to be slightly bitter. But no need to worry: this bitterness will go away in time, of its own accord.

We might summarize this very simple process as follows:

General formula

Water	74 L
Honey	25 L (around 37 kg)
Tartaric acid	50 g
Bismuth subnitrate	10 g
Fresh pollen	50 g

Mead made from honey harvested from fixed-comb hives can be excellent. It ferments very smoothly thanks to the pollen this largely unprocessed honey contains. If you have no pollen on hand while making mead, you can add some fixed-comb honey to extracted honey.*

Proportion of water, honey, etc. needed to produce mead in barrels of various volumes**

Barrel, L	1	2	4	5	10	20	40	50	100
Empty, L	0.01	0.02	0.04	0.05	0.1	0.2	0.4	0.5	1
Spring water, L	0.74	1.48	2.96	3.70	7.4	14.8	29.6	37	74
Honey, L	0.25	0.5	1	1.25	2.5	5	10	12.5	25
Honey, kg	0.37	0.74	1.48	1.85	3.7	7.4	14.8	18.5	37
Tartaric acid, g	0.5	1	2	2.5	5	10	20	25	50
Fresh pollen, g	0.5	1	2	2.5	5	10	20	25	50
Bismuth subnitrate, g	0.1	0.2	0.4	0.5	1	2	4	5	10

Note: 1 L = 1.06 qt = 0.264 gal. 1 kg = 2.2 lb. 1 g = 0.035 oz. Tartaric acid and bismuth subnitrate used in powder form.

---

* Mead makers also commonly use commercially available yeast nutrient from wine-making supply stores. *Ed.*

** This table is excerpted from the following work: *La pratique de l'hydromel sec et liquoreux*, by Mr. Du Chatelle, President of the Société d'Apiculture de l'Est (*Bulletin de la Fédération*, 1896, p. 19).

263. The Guyot glucometer

A glucose meter, or glucometer, is an instrument for measuring the alcohol content that will result when a sugary liquid is transformed into an alcoholic beverage. To measure this content, allow the glucometer to float in a glass containing the sugary liquid, then observe the reading indicated on the instrument, where the water level intersects the *potential alcohol content* scale.

Figure 209 shows a test tube filled with sugary liquid, with the instrument floating in it; if the glucometer reads 14 degrees at the water level *l*, that means that fermenting this liquid would result in an alcohol content of 14%.

So you can use a glucometer to know the alcohol content of your mead in advance. Just let it sink into the mixture of water and honey you've prepared. A glucometer is indispensable for determining how much honey should be added to the water in order to obtain a mead with the desired alcohol content; it's also useful for ameliorating wine with honey (§ 270) and for pyment (grape mead) or cyser—cider mead (§§ 271 and 273).

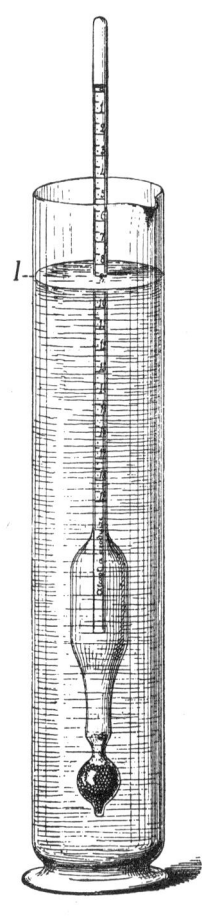

Figure 209. A Guyot glucometer; *l*, liquid level.

264. Using wash water when making mead

After harvesting honey, the honey-covered cappings that fell onto the sieves are thrown into a tub; then you add water and mix everything thoroughly in order to separate all of the honey. As the waxen cappings rise to the surface, you can shape them with your hand into balls that will be used later to produce wax (§ 277).

Next, the water used to wash all of the honey extraction tools is added to this water, already suffused with honey.

This wash water should be used to make mead as soon as possible, since, if the temperature is high and the liquid contains too little sugar, fermentation may set in quickly and render the solution acidic. If you've waited too long, this water will be of no use in producing mead, which could turn into vinegar after fermentation.

When this honey-water is ready, let the glucometer float in it. Let's assume, for example, that it reads 5 degrees; in this case, you should dissolve enough honey in the liquid to cause the glucometer reading to rise to 17 degrees.* Pour the resulting liquid into a barrel. It's unlikely to fill the entire barrel, so you'll need to add some more water, containing the same percentage of honey.

Using a simple rule of three, you can easily determine how much honey and water should be added to the honey-water that has already been poured into the barrel, but hasn't filled it.

For example, if you have a 100-L barrel, and have already poured 35 L of honey-water (reading 17 degrees on the glucometer) into it, then you have 64 L left to fill—since, as we know, approximately 1 L of space should remain empty.

It's easy to calculate how much water and honey should be added to fill the barrel.

If you have 35 L of honey in 100 L of water, you'll need:

$$35/100 \times 64 = 22.4 \text{ L}$$

So you should prepare a solution of 64 - 22.4 = 41.6 L of water and 22.4 L of honey, and add it to the barrel.

265. Time required for fermentation

The duration of the fermentation process will vary depending on the temperature. In spring and summer you should place the barrels outdoors, in the sun; during the winter, place them in a cellar, storeroom, kitchen, etc.—but never in a room where vinegar has been kept.

---

* The number of degrees is never absolutely precise, since all honeys are different; but, in practice, the number given here is enough for identifying the mead's future alcohol content with reasonable accuracy.

Mead is often made after the honey harvest, in order to make use of the honey-water resulting from washing the cappings. In this case, fermentation will take place during the winter, in a cellar, and conclude in the summer, in the sun.*

As fermentation proceeds, the water level will drop a bit inside the barrel. It is often suggested that honey-water should be added bit by bit to keep the level constant, but there is no advantage to doing so.

When you no longer hear the liquid fizzing, fill in the gap all at once with some finished mead, or even with water; then seal the hole with a plug. Now there's nothing further to be done until you're ready to bottle the mead. In any case, don't transfer it from one barrel to another.

Mead produced in the spring can be ready in five or six months, while mead made after the honey harvest and stored in a cellar will require more time to ferment completely.

Generally, mead still appears cloudy when the fermentation is complete—that is, when the glucometer reading is near 0.

A long time must pass from the end of fermentation and the moment until the mead becomes clear—from six months to over a year, depending on the kind of honey used, as well as other factors that have yet to be identified.

It's best to wait for the mead to clear up on its own, since it will only improve with age. Generally, it will clear up more quickly in winter than in summer.

Just as with wines, there are no production methods that can artificially replicate those qualities in mead that come naturally with aging. So there's no reason to be in a rush to bottle it.

To put it simply, a young mead is inferior to a good young wine, while an aged mead is comparable to a respectable old wine.

266. Fining and bottling mead. Barrels and barrel maintenance

As we've seen, once fermentation is complete, it's best to wait for the turbidity to disappear on its own.

---

* The recommendation to place fermenting mead in the sun assumes that you are using a wooden barrel for fermentation. *Ed.*

However, in certain situations, this turbidity can persist indefinitely, and you can attempt to clear up the liquid more rapidly by fining it.

You can clarify the mead using egg whites as a fining agent, just as you would do to wine. After fining, wait until the liquid is perfectly clear before bottling it.

If the mead is still murky even after fining, it's usually because you've fined too early; if this happens, all you can do is resume fining later.

The barrels are placed in a cellar during winter; as we've noted, no mead should be poured into another barrel in any case.

Mead is bottled as usual; but since mead often continues to ferment slightly, it's a good idea to leave the bottles upright for a certain time. If the mead still has a slight sugar content, you could use it to make a sparkling wine. Use champagne bottles (the only bottles that can stand up to high pressure); once they've been corked, using wire, you can store them on their side.

Any barrels can be used in mead production, except for those that have contained vinegar or cider. Of course, the barrels used shouldn't have a bad taste. All too often, in the countryside, barrels in poor condition are used, and the resulting mead has a barrel-like taste that can't be eliminated.

The important thing is to keep your barrels in good condition, indefinitely, as follows: after bottling the mead, first rinse the barrels several times with water, then let them dry out—in a shed, for example—after turning them upside down, with the plughole facing downward, and removing the tap. This will establish an air current between the open plug and the hole in the tap that will dry the barrel. When it's good and dry, burn a sulfur wick inside the barrel, then close the plug and the tap, and store it in this way in a cellar.

267. The hydrometer

There's a very simple instrument, available commercially, called a *hydrometer*, used to determine a liquid's alcohol content. Here is a basic description, which will be more easily grasped if you refer to figure 210.

Let's say you want to determine a mead's alcohol content. Put some mead into a glass, then pass the scaled tube, t, through a hole in a

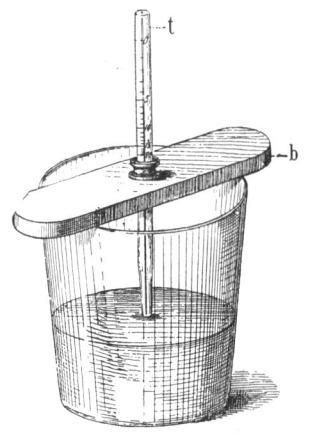

Figure 210. A hydrometer.

small board, *b*. Set the board across the top of the glass, then lower the tube until it barely touches the surface of the mead (the tip of the tube should barely graze the liquid). Now, suck up the mead through the tube, as through a straw, until it touches your lips, then allow the liquid to sink back down; it will stop at some point on the scale marked on the side of the tube. The reading on the scale indicates the alcohol content.

If, for example, the hydrometer stops sinking near the 16 degree mark, counted by reference to the 0 marked at the top of the tube, this means that the mead is 16% alcohol.*

268. Meads with various sugar content

Everything we've said thus far applies to *dry mead*—that is, mead that no longer tastes of sugar at all. You can also make mead that maintains slight sweetness for an extended period; such mead is bottled before it completely loses this hint of sugar. Some prefer this kind of mead to mead that is completely dry.

Sweet mead. Mead that still contains non-fermented honey is referred to as *sweet mead.*

To make sweet mead, add enough honey to water for the glucometer to read 19 or 20 degrees. Since fermentation ends before reaching this alcohol content, the mead will be distinctly sweet.

269. Mead composition

The composition of mead varies, depending not only on the kind of honey it is made from, but also on the way it's produced. Generally,

---

* This device, along with the necessary instructions on how to use it, is available from Mr. Broussard, a manufacturer at 29 Quai de l'Horloge in Paris. The same manufacturer also produces Guyot glucometers.

mead's chemical composition differs from that of wine; it contains some dextrin, less tannin, and less mineral substances—and those it does contain are less alkaline. Moreover, if you analyze mead with a refractometer, the light is refracted to the right by meads, and to the left by wines.

Here, according to Gayon,[*] is an analysis of various meads:

	Meads from the year			
	1886	1887	1889	1891
Alcohol content, %	12.9	13.7	13.4	13.4
Dry substances extracted, g / L	43.75	51.5	46.5	110.5
Glucose and fructose, g / L	12.2	21.27	4.7	72.5
Dextrin, g / L	11.61	8.73	1.9	7.3
Ash, g / L	0.75	0.90	0.10	0.65
Tannins, g / L	0.2		0.23	0.30

270. Ameliorating wine with honey

In regions whose vineyards are near the point where grapevines cannot be cultivated—whether in the north, or at a certain altitude in mountainous regions—the grapes themselves may not have enough sugar to produce a wine with sufficient alcohol content.

This can be known in advance by using a Guyot glucometer (§ 263). If, for example, the glucometer, when dipped into the grape juice, indicates that the wine's alcohol content will be only 5.5%, when the desired content is 10%, you can do the following.

Pour 1 L of grape juice into a container, and dip the glucometer in the juice. If you have 500 g of liquid honey, for example, pour it gradually into the grape juice, mixing thoroughly, until the glucometer reads 10 degrees. When the unused amount of honey is weighed, you find, for example, that 450 g remains; this means that you need to add 50 g of honey per liter to turn a wine at 5.5% to a wine at 10%. The grape juice and the honey should be mixed prior to fermentation.

As a general rule of thumb, 23 g of honey should be added per liter of juice in order to boost the alcohol content of the resulting wine by one percentage point.

---

[*] See *Apiculteur*, 1892, p. 297.

271. Pyment (grape mead)

The term *pyment*, *grape mead*, or *grape melomel* refers to wine to which honey and water have been added prior to fermentation, thus increasing both the overall quantity of the resulting wine and its alcohol content.*

Let's say, for example, that you wish to double your wine production and raise it to 10% alcohol, instead of the 6% the grape must would naturally produce. Add as many liters of water as you have liters of grape juice; the result is double of volume of juice, but now at just 3% alcohol.

Now, add as much honey as needed (23 g of honey for each percentage point you want to boost the alcohol level of the resulting wine—multiplied by the total number of liters).

In our example, we have 10 - 3 = 7, so: 23 g x 7 = 161 g of honey per liter.

This procedure generally yields excellent results; when it's complete, you will have, in essence, made ameliorated wine and mead simultaneously, producing grape mead that reaches the desired alcohol content.**

272. Pomace wines

In the production of pomace (marc) wines, honey can replace sugar. In essence, this amounts to making mead by simply replacing the tartaric acid in a mead recipe with the dregs left after pressing the first wine. The grape skins add tannic acid instead.

---

* *Melomel* refers to any fruit juice fermented with honey to produce a fruit mead, including *pyment* (grape mead), *cyser* (apple mead), and *perry* (pear mead). *Ed.*

** The best way to produce pyment quickly is provided by Mr. Godon (*Apiculteur*, 1896, p. 47). Pour 25 to 30 kg of fresh grapes into a 550-L barrel, open on one end; after crushing the grapes, dissolve honey in water, and pour this honey water onto the grapes. Use 400 g of honey per liter of water for an alcohol content of 16–17%, or 220 to 300 g for 10–12%. Around 50 L of the barrel will remain empty. Re-cover the open end of the barrel with some cloth. Each morning and evening, stir the mixture with a pestle; near the end of the fermentation process, which can last from 10 to 15 days, draw the liquid and pour it back on top.

The total amount of water used will typically equal the quantity of juice from the first pressing.

273. Cyser

In regions where cider is made, it can be ameliorated, resulting in *cyser (cider mead)*, using the same procedure described above in § 271.

274. Honey vinegar

Honey and water can easily be used to make excellent vinegar of trusted origin, and much to be preferred to sometimes tainted liquids sold commercially under the name of "vinegar."

Here's the procedure.

Fill a barrel three-fourths full with a mixture of water and honey, containing 10% honey. Seal the plug hole with some fabric and a stone that will allow air to pass, then place everything in a warm spot or in the sun. Eight or ten months later, the vinegar is ready to consume.

Never put this barrel in a cellar where barrels of wine or mead are stored, since the vinegar's acetic fermentation could spread to the wine and mead and cause them to turn. By the same token, you should never use a barrel that has contained vinegar in the past for storing mead or wine.

You can shorten the time required to produce this vinegar by spreading what is commonly referred to as "mother of vinegar" in the barrel after the peak fermentation.*

As you draw the vinegar from the barrel, replace it with mead diluted to some degree with water.

275. Honey brandy

By distilling mead you can produce an excellent *brandy*—yet another way to obtain a product whose origin you can be certain of. But,

---

* Mother of vinegar, sometimes known as *mycoderma*, is a film-like substance containing bacteria (*Micrococcus aceti*) that turns weak mead into vinegar.

generally, such production isn't possible for commercial purposes, and must be limited to personal consumption, or consumption by a small circle of enthusiasts.

Indeed, to make a liter of brandy with 50% alcohol, you'll need 1.3 kg of honey. Assuming honey is worth 1 franc per kilogram, and keeping in mind the cost of production itself, one liter of brandy will cost 1.60 francs.

The convenient part of all this is that a beekeeper can always find good use for poor-quality honey or otherwise worthless honey residues, to make mead that can be distilled into several liters of good, natural brandy.

276. Uses of honey

There are a great many recipes that call for honey.

For liqueurs, beverages of various kinds, jams, and various dishes—one might almost say, for any kitchen or household recipe—good honey can be used as a sugar substitute.

Honey is also an excellent treatment for coughing and sore throat, as well as for certain stomach ailments. It is used to manufacture many ointments, and low-grade honey is often used in veterinary medicine.

Finally, large amounts of honey—*dark amber honey* in particular (§ 257.1)—are used in gingerbread production.*

277. Wax production

When you wish to extract the wax from old comb, or the wax that results from uncapping comb, without using complicated and costly equipment, you can use the following method.

Install a tap near the bottom of a boiler. Place the boiler atop a tripod that is tall enough to allow a watering can to be easily placed underneath the tap (fig. 211).

Now fill the boiler two-thirds full of water and place it over a fire;

---

* Numerous culinary and medical recipes can be found in the following brochures: *Le miel et son usage*, by Dennler; and *Le miel des abeilles*, by Voirnot.

CHAPTER 19. APIARY PRODUCTS. MEAD

Figure 211. Simplified wax production.

when the water has come to a boil, throw in the combs—then, using a stick, stir the mixture until the wax is entirely melted.*

Once the wax is melted, use the tap to draw the boiling water into the watering can. Using a kitchen colander, ladle up some of the molten mixture of water and wax, and, while holding the colander above the boiler with one hand, use the other to pour all of the boiling water in the watering can into the colander; the water will wash the wax away with it, leaving only some dregs in the colander, which can be discarded. Repeat this procedure until all of the dregs have been removed from the boiler.

Now melt down a new batch of comb, and repeat the same procedure.

When you're done, remove the boiler from the fire, surround it with straw or hay, and cover it somehow to ensure that the wax is purified as it slowly cools. This is the easiest way to produce pure wax.

* Don't add too many combs at once, and turn down the flame once the mixture reaches the boiling point to keep the molten wax from overflowing the boiler, since this wax is flammable. Position the fire such that it only heats the bottom of the boiler.

278. The solar wax melter

Another way to melt wax is to use a *solar wax melter*, sometimes referred to as a *solar wax extractor*.

This is the best procedure to use with wax from uncapped comb; however, old comb is hard to melt using this device.

The solar wax melter consists of a glass-covered compartment resembling an old-fashioned top-opening school desk (*G*, fig. 212) containing a metal mesh atop a tin container. The wax is placed atop this metal mesh (*M*, fig. 212), and when the device is exposed to strong sunlight, the wax melts and is strained through the mesh and into the container.

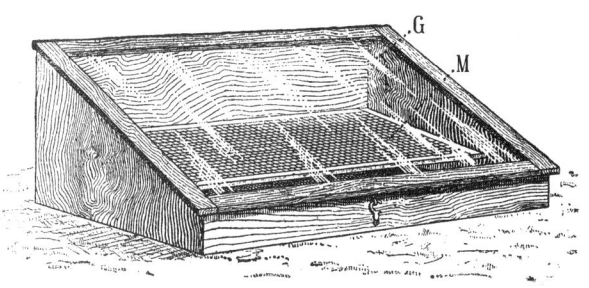

Figure 212. A solar wax melter: *G*, glass pane; *M*, metal cloth.

279. Large-scale wax production

Producing wax on a large scale is an entire art unto itself, and can't possibly be described in this book. It requires a lot of very expensive equipment: a press for extracting wax, a boiler, a purifier, molds, etc. On top of that, the procedures required to produce wax that is perfectly pure are rather complicated and demand a long learning period.

If you happen to have a large amount of wax, it's simpler to sell it to a professional wax maker than to melt it down yourself.

CHAPTER 19. APIARY PRODUCTS. MEAD

280. Making wax foundation

The *Rietsche press*, named after its inventor, is a kind of press that allows you to make sheets of wax foundation yourself.

Using this press allows you to be certain that the wax you use is pure, but it's impossible to produce foundation sheets using this instrument that are as thin as those available in stores, which are made with a machine involving rollers. A certain amount of practice is required before you can get the most out of this device.

281. Telling artificial wax from real wax

Stores often sell artificial wax. Foundation made from artificial wax would lead to major problems. Here's a simple and reasonably reliable way to make sure the wax you're buying is pure.*

Using two small paper tubes, melt, in one, a small stick of beeswax known to be pure—for example, wax taken from a comb built without foundation—and, in the other, a similar small stick of the wax you're testing. Place the two sticks in two bottles or two test tubes, then fill them with benzine (petroleum) and seal them. Pure wax will dissolve very well in petroleum if you shake the tube from time to time, while artificial wax will usually leave behind bits that are undissolved or only partially broken down by the benzine, even if you shake it.

Here's another procedure with results that are less ambiguous, devised by Armand Gaille, a pharmacist. This method is the simplest of all those whose results are certain.

The materials necessary for this analysis are a very small glass funnel, several test tubes that can hold around 50 cubic centimeters (0.5 deciliter) each, some small paper filters, some red litmus paper, a small bottle of liquid ammonia, oil of turpentine, and some 90–95% alcohol. All of this will cost between 2 and 3 francs.

---

* While wax adulteration is a minor concern in the U.S., virtually all wax foundation available commercially today is contaminated with pesticide residue. This prompts many natural beekeepers to give preference to foundationless frames. Foundationless frames also allow bees to build ample drone comb, which is beneficial for the genetic success of the colony. *Ed.*

Three tests should be carried out in sequence.

1. *Test of specific gravity.* Mix one part alcohol and two parts water in a large, ordinary kitchen glass. Then add to this mixture a small piece of wax (about the size of a pea) whose exact weight is known. Then, remove it, and press it several times, still wet, between your fingers and put it back into the liquid. Then, add water gradually, stirring constantly, until the piece of wax floats suspended in the water, neither hitting the bottom nor reaching the surface (or, only doing so at a very low speed). Now take the wax you're analyzing and place it in the liquid after pressing it as described above; if it falls to the bottom of the glass with any speed, or it if rises to the surface with any speed after being submerged, then the wax is clearly artificial. But if the piece of the suspect wax behaves like the pure wax, it may be free of any admixtures—although you will only be sure after the remaining two tests. Indeed, if the manufacturer of artificial wax was careful to use some substances lighter than wax, and others heavier, then the resulting product may well have the same specific weight as the very purest wax.

2. *Dissolution in oil of turpentine.* Place a piece of the suspect wax (about the size of a small hazelnut) into a test tube, then pour in between three and four finger-widths of the turpentine and heat it lightly on an ethanol flame. If it dissolves incompletely, or the solution is extremely murky, or includes sedimentation, then the wax is fake, since turpentine should completely dissolve pure wax.

3. *Chemical testing.* In a glass test tube, boil a *tiny bit of the suspect wax*—a small pea's worth—for several minutes, along with a 1/4 of a deciliter of alcohol (approximately half of the test tube); use an alcohol burner. Then let it cool for a good half-hour at least, then filter it. Now, to the filtered liquid, add an equal amount of water, as well as a small piece of litmus paper that has been turned blue by dipping in ammonia, then half-dried by pressing firmly several times between two sheets of clean blotting paper. Stir the resulting mixture. If after fifteen minutes or so the liquid remains almost clear, and if the litmus paper hasn't returned to its original red color,

then the wax is indeed pure (provided that it has passed the first two tests). If this is not the case, then it's fake. Don't worry if the paper changes color only slightly, or the liquid remains slightly opalescent—this commonly happens even when the wax is pure.*

282. Uses of wax

Wax is used to make oilcloth, wooden floor wax, and wax used in thread, polishes, sculpting, galvanoplasty, printing, candles of various kinds, and certain kinds of ammunition. Finally, wax figures as an ingredient in many pharmaceuticals and chemical substances.

Summary

The honey room
A beekeeper should have a well-sealed, well-ventilated room, or *honey room*, where all of the necessary procedures can be performed away from bees. The room can also be used to store honey and fully drawn frames, as well as reserve honey frames.

Honey
An apiary's main product is honey; wax is a product of secondary importance. Sometimes a beekeeper can profit by selling live bees in fixed-comb hives. Quite often, all three may be sold all at once—that is, hives populated with bees, along with the wax and honey they contain.

There are many kinds of honey, all very different from one another, which crystallize at various rates.

The major varieties of French honey are: Alps or Chamonix honey; sainfoin or Gâtinais honey; honey of the Mediterranean region, one of the finest of which is from Narbonne. The lowest-quality honeys include: heather honey, or honey from Landes; buckwheat honey, or honey from Brittany; blended, the latter two honeys constitute dark amber honey ("red honey" in France). Heather honey is very difficult to extract due to its tendency to stick to comb.

In France, it's easier to sell extracted honey than comb honey.

Mead
In certain regions where honey is produced in large quantities, it is in the beekeeper's interest to make *mead*, for private consumption or even for sale.

---

* *Bulletin de la Société d'Apiculture de l'Aube* (volume 33, 1896, p. 29).

A good mead should have 13–17% alcohol content, and hold up well to mixing with water for drinking.

The best and simplest way to make mead is to mix, before fermentation, one fourth honey by volume with three fourths water, adding a small amount of tartaric acid, bismuth subnitrate, and fresh pollen.

Mead requires a rather long fermentation period, lasting six months or more; but this procedure results in a superior product.

Sweet mead is mead that still contains a certain quantity of non-fermented honey.

Ciders and wines, even pomace wines, can be ameliorated using honey. You can also increase the amount of wine or cider produced by mixing the grape or apple juice with honey and water, which results in pyment (grape mead) or cyser (apple cider mead).

Honey vinegar and brandy

Water containing 10% honey that is left in a barrel for eight to ten months will produce *honey vinegar*.

By distilling mead, you can make an excellent *brandy*, but this process has little commercial potential, and is only pursued by enthusiasts.

Uses of honey and wax

There are many uses for honey: for household recipes, making gingerbread, and in pharmaceuticals and veterinary medicine.

If you want to extract wax on a small scale, you can do so using a boiler or a solar wax melter. The latter device works best for processing wax resulting from uncapping comb.

Processing wax on a larger scale is a special discipline that requires costly equipment and complicated procedures.

There are many uses for wax.

Chapter 20

Bee Diseases and Enemies

283. Foulbrood, or brood rot

Foulbrood is the worst disease that can strike an apiary. In Germany and England, for example, entire apiaries with a large number of colonies have been wiped out in a very short time by this terrible scourge.

Foulbrood is produced by one of those microscopic organisms, now referred to as microbes, which are behind most of the contagious diseases that affect animals and humans. Everyone has heard of a similar disease (called *pébrine*), also caused by microbes, that affects silkworms.

Foulbrood bacteria not only attacks adult bees, but also and especially larvae, and even eggs. This bacteria, *Bacillus alvei*, is shaped like tiny rods (*b*, fig. 213), a few thousandths of a millimeter in length, which divide into easily detachable segments. When these tiny bars are young, they can move, and one can observe these rapid movements under a microscope. When they're older, they become motionless, and when

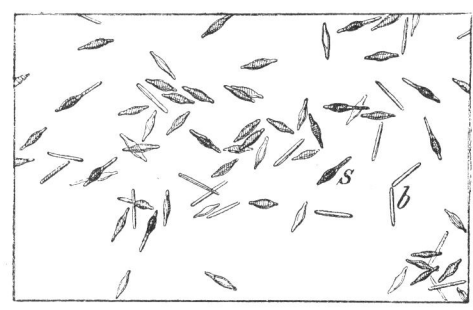

Figure 213. *Foulbrood* bacteria, *Bacillus alvei*, seen under a microscope: *b*, small bar-shaped bacteria; *s*, a spore formed on a bar (drawn from nature).

food in their immediate surroundings becomes scarce, small round formations take shape inside them, which are the seeds, germs or spores (*s*, fig. 213) of this bacteria; and these spores, which can survive a wide range of temperatures, dehydration, and air deprivation, serve to spread this disease.

These spores can stick to bees' bodies, comb, and hive walls, and can even be found in honey or any other substance the bees come in contact with.

And when one of these spores lands in a favorable environment—for example, brood—it germinates like a seed and gives rise to more small bar-shaped bacteria capable of movement, which rapidly divide and multiply in bee larvae, thus causing the disease to spread.

You already have some idea of how quickly foulbrood can spread, and how difficult it is to destroy these microscopic germs, or spores.

284. How to spot foulbrood

1st situation: Foulbrood has recently broken out in a hive. When foulbrood is in its initial developmental period, it's not always easy to spot it, based solely on the colony's outward appearance.

Let's examine some brood comb, paying special attention to the sealed brood. If you notice seals that are neatly pressed inward, or punctured by a small hole, or even torn apart like a torn drumhead (fig. 214), then remove the larva from one of these cells using the head of a pin; if the larva has been turned into a kind of sticky clot, then the brood has been struck by foulbrood.

Meanwhile, if the younger, still unsealed larvae are white and pearly like normal brood, we can conclude that the foulbrood is still in its first stage. Initially, the disease tends to strike larvae that have just been sealed in their cells.

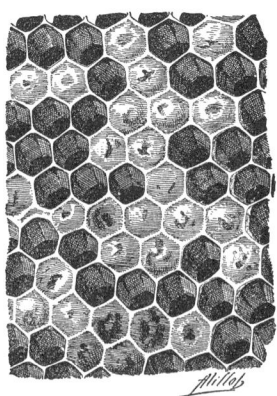

Figure 214. Brood comb fragment attacked by foulbrood.

Another sign of foulbrood at this stage is that the sealed brood is scattered, since, while some bees have emerged from their cells, others remain sealed, decomposing in their cells. A larva that has been infected turns into a sticky mass that sticks to the bottom or sides of the cell; it shows gray or yellow spots at first, and later turns the color of coffee with milk, or brownish. Moreover, the seals of foulbrood-infected cells take on a darker shade that allows them to be spotted right away.

2nd situation: Foulbrood has infected a colony for a long period. In this situation, you can often tell that the colony has been struck by foulbrood even from the outside, due to the sluggish activity of the bees and the odor of decay emanating from the hive entrance.

When you examine some brood comb, you notice that not only the sealed brood, but even the young larvae have been infected—the larvae are yellowish or brownish in color, and elongated in their cell instead of assuming their usual curved shape.

If, as the infected colony grows weaker and weaker, it is robbed, then the robber bees can spread the infection throughout the apiary.

285. Hygienic measures for preventing foulbrood

We're not entirely sure how exactly foulbrood strikes bee colonies. Still, it has been proven that certain circumstances can make a colony vulnerable to infection, and they must be avoided at all costs in your hives.

1. *The brood must not be left uncovered.* Throughout all of your procedures, such as moving a hive, spring feeding, creating small nursery hives for raising queens, etc., it's very important to proceed prudently and at the proper time, as indicated previously in this book; the bees should be enough in number and clustered in such a way as to always cover the colony's brood. If the brood grows cool while abandoned, even if briefly, by the bees, it risks contracting foulbrood.

2. *Always take the greatest possible precautions against robbing.* We've already explained why it's important to prevent robbing. Now we can

add another reason: any attempt at robbing by bees from a neighboring apiary, over which you have no control, could introduce foulbrood into your apiary.

3. *Melt down pieces of comb containing brood when transferring or dissolving a hive.* When you're directly transferring a hive, don't simply throw away drone comb or bits of worker brood comb that you can't use. Melt it down along with the wax.

286. Treating foulbrood

It isn't always easy to treat foulbrood, and once the infection has reached its most serious stage, the simplest and wisest thing to do is eliminate the entire colony. You'll learn more below (§ 287) about how to disinfect the hive and the comb.

Many remedies for foulbrood have been proposed. Hilbert was one of the first to advise using antiseptics, especially salicylic acid. Without delving into various methods that have been tried, to varying degrees of success, to counter this terrible infection, we'll restrict ourselves to describing the procedure that seems the most likely to succeed.*

If, during your spring inspection, you notice that a colony is showing signs of first-stage foulbrood, move all of its bees into a new hive containing frames primed with foundation or pieces of comb—thus, so to speak, transforming the colony back into a swarm. This procedure should take place approximately three weeks before the main honey flow. Be sure to put a few mothballs in a small cloth sack and place it at the end of the hive opposite that in which the bees are clustered.

Also, prepare the following solution ahead of time:

Melt 2 lb (1 kg) of sugar in 1 qt (1 L) of hot water, then add 10 g of 12% solution of salicylic acid in alcohol (the former can be found in pharmacies). Every two or three days, give the colony a pint (0.5 L) of this syrup, over a period of three or four weeks.

---

* Treatment with salicylic acid yields mixed results. Burning all affected colonies and equipment is regarded as the most reliable method for stopping the spread of this highly contagious disease. *Ed.*

If you notice signs that a colony is infected during your fall inspection, simply put some mothballs in it as described above, and wait for the following spring to return it to a swarm state and treat it with salicylic acid.

In any case, even if there are no signs of foulbrood, it's a good idea to *always keep some mothballs in your hives*—that is, one or two of them, in a small cloth sack, for each hive.*

287. Disinfecting a hive infected by foulbrood

It's very important to disinfect an infected hive as quickly as possible after you've dissolved it or returned it to a swarm state.

Run the comb containing honey through the extractor; this honey can be used to make mead, but should in no case be given to bees. The comb should be melted down, and the frames submerged in boiling water, or soaked in a 10% solution of sulfuric acid. This same solution can be used to wash thoroughly all the parts of the hive; finally, burn some sulfur in a room with the hive and the frames inside.**

288. Dysentery

This illness usually breaks out in winter, and sometimes in autumn; it's a kind of indigestion that can be spotted by the buildup of excrement in the hive, which will give off a foul odor.

Dysentery is usually due to a prolonged wintering period in humid, insufficiently ventilated air. So hive ventilation during winter, which we've recommended elsewhere, is important in this regard as well.

This sickness can also be caused by giving the bees excessively watery feed, or by the fact that the bees didn't have enough time at the

---

* Mothballs (naphthalene) are toxic and should not be used inside beehives. Naphthalene builds up in wax and contaminates honey. *Ed.*

** Boiling water is not hot enough to kill foulbrood spores. Other disinfection methods, such as scorching, are not fully reliable either. One efficient method is to immerse woodenware in paraffin wax at 320°F (160°C) for 10 minutes, but it is only practicable for a sizable operation. All in all, burning all affected equipment is usually the most reliable and economical method to prevent reinfection. *Ed.*

end of foraging season to drive off the excess water contained in the most recently gathered nectar, or in the syrup you gave them. Yet another reason to avoid feeding in the fall.

Experience has shown that dysentery strikes Italian bees and Italian hybrids more often than ordinary black bees—just one more reason to avoid keeping foreign races, as we've suggested elsewhere.

All in all, if you're keeping black bees, then providing carefully for the wintering period and avoiding fall feeding will ensure that dysentery is a rare occurrence in your hive.

If, however, you find that a colony has been struck by this disease, replace its bottom board, then put back the hive, propped up on some wedges.

Generally, dysentery is not a serious illness, and often vanishes of its own accord in spring.

289. Other bee diseases

Bees are prone to several other diseases, exceedingly rare or little studied.

Vertigo causes bees to spin out of control in flight, fall, and die; this disorder has been attributed to honey made from certain kinds of flowers.

The nectar of some flowers is sometimes so poisonous that bees will die on the spot the moment they absorb it. This malady is referred to as *narcotism.*

Sometimes, larvae and pupae will dry out in their cells, without changing color; this is called *chalkbrood.* The bees will often remove this dried-out brood themselves.

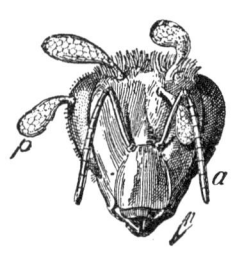

Figure 215. A bee's head with orchid pollen, *p*, stuck to it; *a*, antennae (enlarged).

Often, viscous formations stuck to bees' heads, much like plumes in a cap (*p*, fig. 215), are erroneously believed to be a disease (*swelling of the antennae*). These formations are simply balls of pollen from the stamens of several varieties of orchid.

290. The greater wax moth (*Galleria mellonella*)

The moths referred to as *greater wax moths* or *waxworms* are the only insects that pose a serious danger to bees. There are two major varieties: the larger (fig. 216) is the most common in the North of France, while the smaller one is more often found in southern regions (fig. 218).*

Figure 216. Wax moth (*Galleria mellonella*, actual size).

Wax moths can lay their eggs on flowers, leading bees to gather them along with pollen or nectar and introduce them into the colony themselves. The adult moth can also get into a hive directly and lay eggs there. When the egg hatches, a larva, or caterpillar, emerges; it has sixteen very short legs (fig. 217). These caterpillars are extremely active, twisting around like tiny worms and digging into wax, which is their primary source of food, and where they build long tunnel-like formations (*t*, fig. 219) whose interiors are lined with silk—especially in those sections of the hive not occupied by

Figure 217. Wax moth caterpillar (*Galleria mellonella*, actual size).

Figure 218. Lesser wax moth (*Achroia grisella*, enlarged).

the bees. They don't eat honey, but if the tunnels they dig are numerous, the comb structure may be seriously compromised, and the queen's egg-laying process will be disrupted.

After a certain period, the caterpillars turn into chrysalises surrounded by white cocoons, grouped alongside one another (fig. 220).

Adult insects later emerge from the cocoons: grayish moths whose shape varies depending on the variety of wax moth (figs. 216 and 218). In winter, the caterpillars remain numb, regardless of their age;

---

* The two species are lepidopterans; the larger is *Galleria mellonella*, or greater wax moth, and the smaller is *Achroia grisella*, or lesser wax moth.

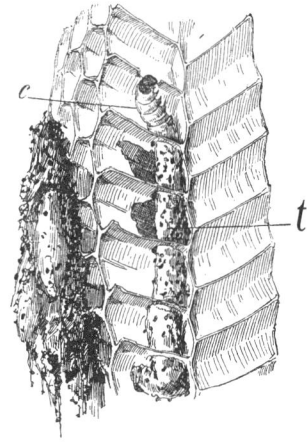 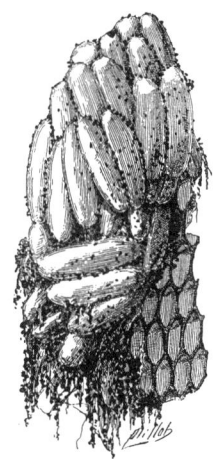

Figure 219. Tunnels dug through comb by the wax moth: *c*, caterpillar; *t*, tube (actual size).

Figure 220. Cluster of wax moth cocoons in the comb of an infested hive (actual size).

it's usually in spring, during the first warm spell, that the wax moth's activity ramps up. There are usually at least two generations of wax moth per foraging season.

291. How bees control the wax moth

If a colony is strong and well organized, it has no reason to fear the wax moth. Indeed, bees are constantly destroying wax moth larvae, cutting holes in the infested comb to drive out the caterpillars, which they kill, then discard outside the hive, as one may often notice in front of the hive.

So only orphaned or excessively weakened hives are threatened by the wax moth, since the bees aren't active enough to fight against moth invasions. Of course, if you sulfurize any combs you harvest before putting them back into your hives, as described in § 86, you'll be taking the most effective step against these invaders.*

---

* Paradichlorobenzene (active ingredient in modern mothballs) is currently approved in the U.S. for preventing wax moth damage to stored comb. Freezing empty comb and storing it in a cold well-ventilated space is a good chemical-free alternative. Naphthalene should never be used to protect comb as it builds up in the wax and contaminates honey. *Ed.*

If a weak colony suffers a major attack by wax moths, then leave only brood comb in the hive, and add some comb that has been sulfurized.

Dissolve any queenless hives—which almost always end up being attacked—as described previously.

292. Other enemies of bees

1. Insects. Another large-bodied moth, the *death's-head hawkmoth*,* breaks into hives (fig. 221) to steal honey—it can escape with up to 2 oz (60 g) at a time. Inside the hive, this moth isn't afraid of being stung, and the bees attempt to block its entry using walls of propolis.

The caterpillar of the death's-head hawkmoth feeds on potatoes, but the belief that it was introduced to France from North America along with the potato is mistaken; the moth was long known in Europe prior to the potato's introduction. Its caterpillars can also feed on wild

Figure 221. A death's-head moth attempting to enter a hive.

---

* Also called the death's-head sphinx (*Acherontia atropos*). (These hawkmoths are native to Eurasia and are not a concern in the U.S. *Ed.*)

plants of the nightshade family—for example, on bittersweet nightshade (*Solanum dulcamara*), black nightshade (*Solanum nigrum*), etc.

Wasps, hornets, large *dragonflies,* and the hymenopteran known as the *beewolf* (fig. 222) are carnivorous insects that attack bees. You can often see them flying amidst bees, trying to grab one and make away with it, when the bees are foraging in large numbers around honey plants. When a beewolf takes a bee, it paralyzes it with its sting, then moves it to a hole it has dug in the ground, already containing an egg; when the egg hatches, the larva will feed on the bee (fig. 222).

Figure 222. A beewolf carrying a bee to feed its larva (actual size).

Figure 223. *Meloe* beetle larva (triungulin, enlarged).

In their larval stage,* *oil beetles*, or *Meloe* (fig. 223), will take up a position in a nectar-bearing flower to wait there for wild honey-producing insects, to which they attach themselves, sometimes penetrating the bee all the way up to their mandibles, and hitch a ride back into the colony, where they complete their growth process by eating honey. Domestic bees are often attacked by oil beetles, and try to break free of their grasp by violent movements, usually without success. Some bees may even die during these convulsions, which are sometimes mistaken for a kind of disease (the *May sickness*, for example). Any *Meloe* beetles that the bees bring back into the hive are thrown out by other bees and prevented from establishing themselves in the hive.

The *Braula*** or *bee louse* is a parasite—fairly large, relative to the

---

* These larvae are referred to as *triungulins*, a term that includes several species of beetles, of the Cantharidin group.

size of the bee (about the size of the head of a pin) and reddish-brown in color—that lives attached to a bee's hairs, without causing the bee any obvious harm.

The bee beetle, or *Trichodes apiarius*,*** is a greenish-blue insect whose elytra are black with red stripes; it may take up residence in comb, especially when the comb is humid. However, this insect doesn't cause any significant damage.

Ants are more of a nuisance than a real danger. Smaller species of ants sometimes appear beneath a hive's top, due especially to the heat emanating from the bee cluster.

Figure 224. The bee beetle (actual size).

2. Spiders. Spiders often trap bees in their webs, and can be especially harmful to covered apiaries that lack sufficient upkeep.

The *Trichodactylus***** is a small parasite that is often found on bees, attached with its curved claws. It simply uses the bees as a form of transportation, without doing them any harm.

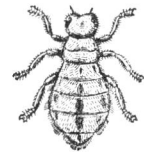

Figure 225. *Braula caeca* (bee louse, enlarged).

3. Reptiles, birds, and mammals. Some bees fall prey to *lizards, toads* and certain *insectivorous birds*.

Field mice, mice and *badgers* are more dangerous enemies. Field mice are highly common everywhere, and often break into hives, even when very small, where they eat everything in sight and, often, build nests. We've already mentioned (§ 76) some precautions you can take against such attacks.

** *Braula caeca.*

*** *Clerus apiarius.*

**** *Trichodactylus* or *Chaetodactylus*, Acari (mites).

Badgers have been known to overturn hives to eat their honey, of which they are very fond.

293. Plants that are harmful to bees

The flowers of *milkweed** and various other species of *Asclepias*, which are often grown in gardens, can cling to bees' legs, holding them until they die (fig. 226). When these plants flower, the ground around them may be scattered with a considerable number of bee corpses, accumulated as the bees are drawn to the milkweed nectar, only to die in their attempt to harvest it.

Figure 226. Bees stuck in the milkweed flowers: *a*, a bee at the moment it is first caught in the flower; a pollen ball is seen on one leg; *b*, a dead bee, caught by a hind leg.

Figure 227. Bees caught in the flowers of the *hooked bristlegrass*: *a*, a bee landing on bristlegrass; *b*, a bee stuck in the bristlegrass; *c*, section of a barb magnified to show the inverse barbs that snag the bees and hold them fast.

* *Asclepias cornuti.*

CHAPTER 20. BEE DISEASES

There are other plants that produce no nectar but can still harm bees when located close to hives. These include certain grasses with inverse barbs (*c*, fig. 227). For example, bees sometimes become stuck in *hooked bristlegrass*.* So you should refrain from planting milkweed in your gardens, and pull up any bristlegrass that is near your hives.

Summary

Bee diseases

The most fearsome bee disease is foulbrood, which can infect entire apiaries, and is spread by the bees themselves.

A general precaution against this disease is to always keep a few mothballs in each hive.** You can help prevent outbreaks by avoiding spring feeding, and by working your bees as little as possible. If foulbrood is still in its early stages, you can try to heal your colony by converting the hive back to a swarm state and treating it with salicylic acid.***

Dysentery is a much less serious illness; it is most common in late winter, and often disappears on its own. It can usually be avoided by ensuring that your hives are well ventilated during wintering.

Enemies of bees

Wax moths only pose a serious threat to extremely weak or queenless hives. Their attacks rarely amount to much, provided that you sulfurize your comb after harvest.

The only remaining enemy to be feared is the field mouse; you can protect your hives against attacks using screens made of perforated sheet metal.****

---

* *Setaria verticillata*.

** The use of mothballs (naphthalene) is no longer recommended as this toxic chemical accumulates in wax and contaminates honey. *Ed.*

***Salicylic acid is not fully reliable for treating American foulbrood. Burning all affected colonies and equipment is regarded as the best method for stopping the spread of foulbrood. *Ed.*

**** This book was written before Varroa mites were present in Europe and the Americas. The use of locally adapted disease-resistant bees as recommended in this book is of critical importance for the long-term success of treatment-free beekeeping. Many management approaches described here (regular breaks in the brood cycle through natural or artificial swarming, leaving bees ample honey reserves for the winter, maintaining strong colonies in stationary apiaries, increasing spacing between hives, infrequent inspections, etc.) are further helpful for the control of Varroa mites and other pests. *Ed.*

Chapter 21
Nectar and Nectar Glands

294. Nectar glands*

Nectar, the sweet liquid that is the primary source of bee honey, is secreted from the surface of certain parts of the plant that are usually found on the inside of the flower, and near the flower's base.

Vaillant** once called the parts of the flower that produce sugary substances honey glands (*mielliers* in French), but today they're called nectar glands or *nectaries*, and a distinction is made between floral nectaries, which are part of the flower itself, and extrafloral nectaries, which can be found on other plant organs (this latter kind of gland is far less frequent).

295. Sugars contained in nectar glands

Sugars always build up in the tissues near a flower's base; this stockpile of sugar is a kind of reserve that the plant draws on, after blooming, for the initial development of fruit and seeds. But the mere presence of nectar glands doesn't always mean there is nectar. Indeed, the sugary liquid only reaches the plant's exterior when conditions favor the transpiration process by which the liquid is secreted.

A given plant may produce nectar in one region, and no nectar at all in another part of the country. There are also plants that never

---

* This chapter is authored by Gaston Bonnier, based on *Les nectaires* (Annales des sciences naturelles, 1879), as well as inedited observations.

** *Discours sur la structure des fleurs*, 1717.

produce nectar under any circumstances, despite having sugar-rich tissues at the base of their flowers.

Nectar develops when a plant exudes water, which, passing from the roots and up through the plant, carries with it some of the sugars contained in the nectar-gland tissues.

These sugars come in two forms: *sucrose*, which is similar to ordinary sugar (cane sugar or beet sugar), and *glucose*, similar to fruit sugar—like, for example, the fine white powder seen on prunes.

Nectar itself, which is primarily composed of water in which these sugars are dissolved, contains a mixture of cane sugar, glucose, and fructose.

The tables below give a sense of what nectar is made of.

HONEYSUCKLE NECTAR *(Lonicera periclymenum)*

Water	76
Sucrose (cane sugar)	12
Glucose and fructose (fruit sugar)	9
Dextrin, gums, minerals, and loss	3
Total	100

LAVENDER NECTAR *(Lavandula vera)*

Water	80
Sucrose (cane sugar)	8
Glucose and fructose (fruit sugar)	7.5
Gums, residues, and loss	4.5
Total	100

The percentage of sugar found in nectar varies widely depending on the flower—to such an extent that some flowers have highly developed nectaries, and large quantities of nectar, but bees are never seen collecting it.

For example, everyone knows the fritillary, or crown imperial—a beautiful spring-blooming garden plant. Its flower contains six nectar glands that, in bloom, secrete six large drops of nectar. Why don't bees gather it? Because this nectar doesn't contain enough sugar, as the following table shows:

FRITILLARY NECTAR *(Fritillaria imperialis)*

Water	95
Sucrose (cane sugar)	1
Glucose and fructose (fruit sugar)	1.5
Gums, residues, and loss	2.5
Total	100

Now we can understand why the bees don't bother to collect this nectar, which contains just 2.5% sugar.

On the other hand, you may sometimes find (using a large magnifying glass) extremely small droplets of sugary liquid on a flower—or you may not even see any at all. And yet bees visit these flowers to gather a sugary substance. This is the case with gorse (*Ulex europaeus*) or anemone (*Anemone nemorosa*). These flowers have no visible nectar, yet bees do indeed visit them in the spring—and not only for their pollen.

If we look more closely, we notice that the bees use their proboscis to draw a kind of highly concentrated, sugary juice from the bottom of these flowers—a juice that just barely oozes to the surface. The bees will even draw from the interior of the nectar-producing tissue, which, in these plants, is soft and spongy. If, in such cases, you'd like to determine the composition of this sugary juice, you can collect it from the crop of bees who have just visited such flowers exclusively. Were you to do so, you'd see that the liquid the bees extract from these flowers is extremely rich in sugar. In fact, its sugar content can exceed 65%.

This explains why the bees forage so diligently among these flowers: here, they can gather a highly concentrated syrup, while all they can get from other flowers—albeit with less work—is essentially sugar water.

296. Nectar contains much more water than honey does

If we compare the compositions of honey provided earlier (§ 257.2) to those of nectar that we've just seen, we notice that there is generally much more water in nectar than in honey.

Nectar contains 70–80% water, while honey only contains 20–25%.* So before capping honey, the bees must spread it out across the comb to allow the amount of water that makes up this difference to evaporate—namely, water weighing approximately 1.5 times the weight of the honey itself. So the volume of the collected nectar falls by roughly three-fourths by the time it is transformed into capped honey. This explains the high humidity found in the hive during intense honey flow, and not to mention the number of ventilator bees, which rises along with the amount of nectar collected, since these bees produce the air circulation that promotes evaporation.

This important observation also shows us why you should leave the bees such a large amount of empty comb: to allow them to spread out their honey. They only put a small amount in each cell to allow the water to evaporate, until the honey contains the desired proportion and can be capped.

But this desired proportion depends not only on the way in which the bees cause nectar to evaporate, but also on the temperature outside the hive. If, for example, the nectar is collected very late in the season—like the nectar of ivy flowers, for example—in a year with early cold spells, then the bees may be forced to cap it when its water percentage is still a bit higher than usual. Sometimes, at the end of the season, when the temperature is too low, you may even see some honey, spread about the cells, that the bees have simply given up on capping, because the temperature is too low to allow the water to evaporate from it.

The upshot is that all the attempts to calculate an ideal hive capacity based on the space left for brood in comparison to that left for honey reserves are always incorrect, since such calculations fail to take into account the space the bees require for temporarily spreading out

---

* Honey water content may be as low as 17% or even less. Honey with over 19% moisture may ferment. *Ed.*

their nectar. Even if we factor this in, it's still very difficult to arrive at accurate results, since the rate of nectar evaporation depends on the season and the outside temperature.

297. Honey has a different composition than nectar

A second very important observation emerges when we compare the composition of honey to that of nectar: honey tends to contain more glucose and less sucrose than nectar. And this is always the case when we compare a given batch of honey to the nectar it was made from. For example, pure sainfoin honey contains less sugar than sainfoin nectar; heather honey contains less sugar than heather nectar, etc. Why? Because the nectar undergoes a more or less complete transformation while in the bee's digestive tract, due to the effect of a special substance* that is found there.

This substance partially transforms sucrose into glucose and fructose. Consider the following examples:

SAINFOIN *(Onobrychis sativa)*
Percentage of sugary substance

	Nectar	Honey
Sucrose	57.2	8.2
Glucose and fructose	42.8	91.8
Total	100	100

As we can see, once the bees have transformed sainfoin nectar into sainfoin honey, the amount of glucose and fructose rises to about ten times the amount of sucrose.

It's important to note that these studies were carried out in the same location (in Louye, in the department of Eure) using nectar and honey gathered in the same sainfoin field.

So if a given kind of nectar contains much more cane sugar than the previous example (and this is the case at high altitudes, in mountainous regions—for example, with certain crucifers that grow in the

---

* This substance, called *invertase*, is produced by the bee's digestive tract, and has the property of transforming sucrose into two simple sugars (glucose and fructose).

mountains)—then the resulting honey contains a much larger proportion of cane sugar than the previous example.

If, on the other hand, the nectar only contains a small percentage of cane sugar, like heather nectar, it will produce a honey that contains almost nothing but glucose and fructose.

298. Nectar glands outside the flower

Sometimes nectar glands are located outside the flower. From a beekeeping standpoint, the most important such glands include those found at the base of the flowers of the common vetch (*Vicia sativa*); these glands are found on small, special leaflets shaped like arrowheads, which are called stipules (*s*, fig. 228). Figure 228 shows one of these stipules, where the surface of the nectar gland, shaded a darker gray, is marked *n*; there, we can see a drop of nectar forming, *d*. Bees gather this extremely sugar-rich liquid even when the plant has yet to produce flowers.

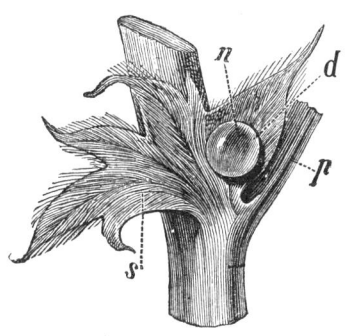

Figure 228. The stipule of the common vetch (*Vicia sativa*): *p*, petiole of the leaf; *s*, stipule; *d*, drop of nectar, through which the dark patch of the nectary, *n*, is visible.

Quite often, one also finds nectar glands in the form of slight protuberances at the base of a leaf stem; under certain circumstances, these glands can secrete a sugary liquid that bees will collect. This is the case with the leaves of cherry trees, plum trees, and hawthorn.

The nectar glands on the leaves of the castor-bean plant are highly developed, and you can even find them on the newly-sprouted plant's cotyledons (fig. 229). These glands produce a liquid that is rich in sugar.

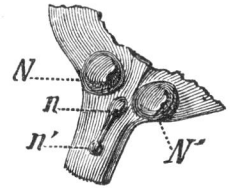

Figure 229. Nectaries at the base of the cotyledon of the castor bean: *N*, *N'*, large nectaries; *n*, *n'*, small nectaries.

In other cases, there are specially-shaped leaves, called bracts, that are found near flowers and produce nectar, as with certain varieties of leadwort and certain species of knapweed.

299. The nectar glands of the nasturtium, hellebore, and horse chestnut

First, let's look at some flowers whose nectar glands are located in the calix (the cup formed by sepals) or in the corolla (the cup formed by petals).

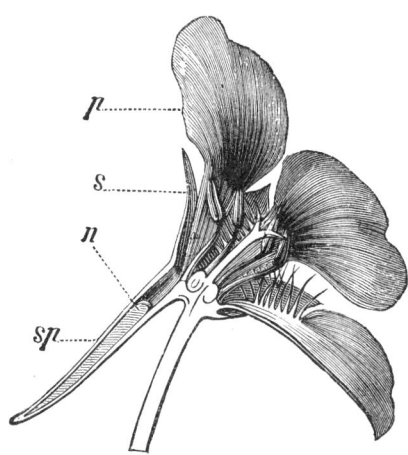

Figure 230. Flower of the garden nasturtium, lengthwise cross-section: p, petal; s, sepal, forming a nectar-bearing spur, sp; n, nectar contained inside the spur.

1. Nasturtium (Tropaeolum majus). Sugars are stored in a kind of spur (sp, fig. 230) found at the base of the calyx. When nectar is abundant and fills the spur, bees can gain access through the inside of the flower itself, but when it is only found at the bottom of the spur, it is too far from the flower's interior to be reached in this fashion. However, bumblebees, whose mandibles are stronger than those of honey bees, often cut through this spur from the outside in order to extract the nectar. In such cases, honey bees may take advantage of holes already cut by bumblebees to gather the liquid from the outside if they are unable to reach it from the inside.

2. Hellebores (Helleborus foetidue, viridis and *niger).* Fetid hellebore, green hellebore, and black hellebore (or Christmas rose) have numerous petals that are completely transformed into nectar-bearing cups.

These cups (fig. 231) are often filled almost to the brim with sugary liquid, even when the temperature outside reaches several degrees below zero. Since hellebores bloom in winter, these plants can serve as a resource to bees who leave the hive to forage for nectar even during wintering.

Wolfsbane and columbine also have petals with nectar-bearing spurs. Wolfsbane flowers are often punctured by bumblebees while still in bud form, leaving behind nectar for bees to come harvest later. Columbine petals extend back through the exterior of the flower in the form of a kind of curved spur, with a rather soft consistency; one can sometimes see not only bumblebees, but even bees tearing at these spurs with their mandibles in order to reach the nectar.

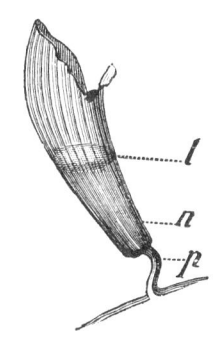

Figure 231. Black hellebore petal with a nectar-bearing cup: p, petiole; n, base of the petal with cup (nectar gland); l, the level of nectar, which you can see through the semi-transparent petal.

3. Horse chestnut (Aesculus hippocastanum). In spring, horse chestnut flowers serve as an abundant source of nectar. Their nectary consists of a kind of lip located inside the sepals and petals; when the flower is spread wide open, bees can easily gather the nectar, which contains certain acids and a substance called aesculin. The resulting honey is rather poor in taste, but this is usually of little concern, since the bees use up this honey—produced in early spring—to feed their brood.

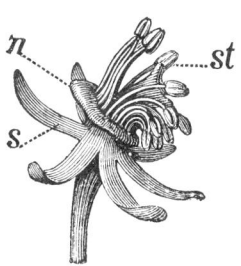

Figure 232. Reseda flower with petals removed: n, nectar gland; st, stamens; s, sepals.

300. The nectar glands of reseda, violets, peach trees, and legumes

Nectaries can be found in an extension of the stamens, or at their base. Let's look at a few examples.

1. *Reseda.* Wild resedas (*R. lutea*, *R. luteola* and *R. phyteuma*) and cultivated resedas (*R. odorata*) have honey-producing flowers. Inside

the flower, the cluster of stamens swells into a kind of prominent reddish disc, on whose surface a nectary (*n*, fig. 232) secretes a sugary liquid that bees can easily reach.

2. *Violets.* Two stamens on a violet flower have extensions on their back that lead into a spur formed by the base of a petal. In this flower, it is the extensions that produce nectar, which drips down and gathers in the bottom of the petal's spur.

Generally, bees are unable to reach it, but they will collect it when the spur has been punctured from the outside by a bumblebee.

3. *Peach trees.* In peach, almond and apricot trees, the nectar-producing tissues form a kind of cup inside the flower, around its entire circumference, beneath the stamens.

These fruit trees are very rich in nectar, and the honey they produce has an excellent taste.

Bees don't always wait for these flowers to open before attempting to reach the sugary liquid they contain. As with many other melliferous flowers, bees will use their mandibles to push aside the still-folded petals, forcing the flower open in order to collect its nectar.

4. *Legumes, or Papilionaceae.* The legume or Papilionaceae family includes flowers that are rich in nectar.

For example: black locust (*Robinia pseudoacacia*), sainfoin (*Onobrychis sativa*), white clover (*Trifolium repens*), black medic (*Medicago lupulina*), etc.

These plants' nectar-producing tissue is found at the bottom of the flower; it forms a kind of thick fold (sometimes with a tongue-like extension), and at times can produce so much nectar that the entire interior of the flower can be full of it, to one depth or another. In certain varieties of these flowers, such as the black locust and sainfoin, the petals are spread widely enough to allow the bee to stick its head in and reach the nectar. In others, like the white clover and black medic, the tube formed by the petals is very narrow, but it's shallow enough for a bee to suck out the nectar by extending its proboscis.

In red clover, the tube is narrow and deep, so bees are typically unable to collect from this plant unless the honey flow is extremely plentiful.

In other legumes, like beans (including green beans), the flowers are large, but the bees have a hard time reaching the nectar, and can usually only reach it through holes already cut into the flowers by bumblebees.

301. The nectar glands of crucifers, anemones, heathers, and buckwheat

1. *Crucifers.* The crucifer family includes many highly melliferous plants: wild cabbage (*Brassica oleracea*), cabbage (*Brassica napus*), woad (*Isatis tinctoria*), etc.

The nectaries of crucifers are located at the base of the stamens (*n*, fig. 233), sometimes even ringing them completely, in the form of small protuberances of slightly variable shape.

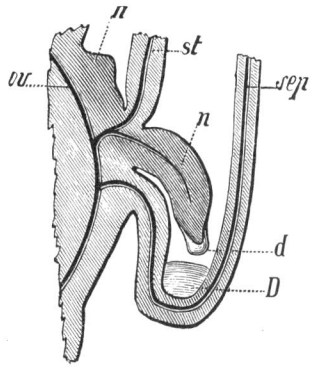

Figure 233. Magnified, partial lengthwise cross-section of a Brassicaceae flower: *n*, nectar glands; *d*, small drop of nectar falling from the nectar gland to combine, at *D*, with existing nectar that has collected along the curved bottom of the sepal, *sep*; *st*, cross section of a stamen; *ov*, part of the ovary (the sections where sap accumulates are shaded).

It's a rather curious spectacle to see bees visit cabbage or wild cabbage flowers, for example, since depending on how abundant the nectar is, they may collect it in various ways:

1. From the flower's interior, by inserting their proboscis in between the stamens and petals.
2. When the nectar is plentiful, from the exterior, by placing the proboscis in the gap between two sepals.
3. When the nectar is extremely plentiful, from the side, by placing the proboscis between a petal and a sepal.

Sometimes they can even harvest nectar from underneath the flower—for example, one may observe foraging bees collecting, in large quantities, the nectar produced by the peduncle beneath the flower of arugula (*Eruca sativa*).

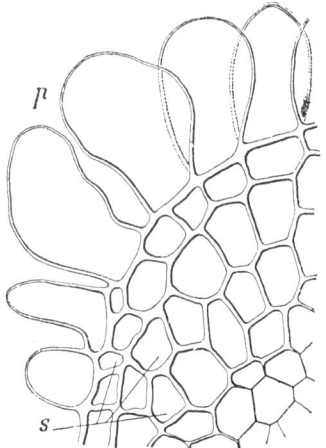

Figure 234. Portion of tissue at the base of the stamens of the wood anemone, magnified cross-section; *s*, sugar cell of the nectar gland; *p*, papillae filled with sugary liquid.

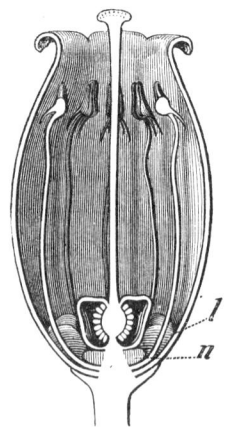

Figure 235. Flower of bell heather, lengthwise cross-section: *n*, nectary; *l*, nectar.

2. *Anemones.* As we've seen above, bees will sometimes extract a highly sugary liquid from some nectar glands that don't produce visible droplets on their exterior. This is the case with anemones, a portion of whose nectar-producing tissues are shown in figure 234. On their surface, we see some papillae, *p*, through which a thin layer of nectar can seep. During spring, a bee can stick its proboscis through these papillae to reach the nectar.

3. *Heathers.* With heathers, the nectar gland consists of a prominent bulbous structure inside the flower, at the base of the stamens (*n*, fig. 235).

The amount of nectar heather produces can vary widely depending on exterior circumstances. There may be no liquid to be found on the surface of this structure, or, quite the opposite, you may find nectar in great abundance, filling the entire bottom of the flower (*l*, fig. 235).

Bees always forage on the flowers of common heather (*Calluna vulgaris*) from the inside. As for the flowers of other kinds of heather (*Erica*), if the corolla has not been punctured by a bumblebee, then bees forage from the interior; but if it has been, they prefer the holes cut by the bumblebees, since the work goes more quickly that way.

4. *Buckwheat (Polygonum fagopyrum)*. Buckwheat nectar glands consist of small, round masses (*n*, fig. 236) found at the base of the stamens, similar to those of the cabbage or wild cabbage.

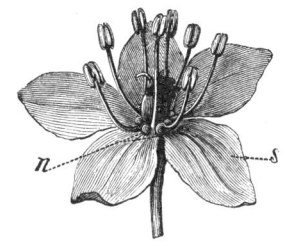

Figure 236. Buckwheat flower: *s*, a sepal; *n*, a nectary.

302. The nectar glands of periwinkles (*Vinca*), mints (*Lamiaceae*), figworts (*Scrophularia*), and houseleeks (*Sempervivum*)

Nectaries may also extend from the pistil—that is, the organ in the middle of the flower where the seeds form. Here are several examples:

1. *Periwinkles (Vinca major, Vinca minor)*. A periwinkle flower features two fleshy, yellowish masses that are located right up against the flower's pistil, and are larger than it is—these are the nectaries. Figure 237 shows a cross-section of this entire complex—the pistil and the nectar glands—clearly showing how prominent the sugary tissues are in this flower.

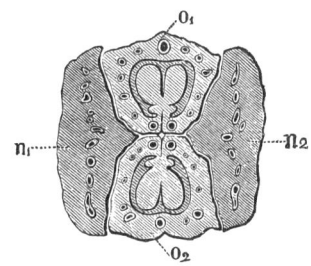

Figure 237. Magnified cross-section of the ovary and nectaries of the periwinkle: n_1, n_2, nectaries; o_1, o_2, ovary.

2. *Mints (Lamiaceae)*. Generally, the plants in the Lamiaceae family are rich in nectar, and the honey they yield is highly aromatic, containing—at least in small quantities—the perfumed essences produced by the plants in this family.

The sage flower, for example, features four whitish protuberances at the base of the pistil. These nectaries are unequal in size (the one near the front of the flower is much larger than the others), but all four emit a sugary liquid, and during heavy honey flow you may see the four droplets combine into one and fill the entire bottom of the flower. If we bisect the flower lengthwise and examine it under a magnifying glass, we can see how prominent the nectar glands are (*n*, fig. 238).

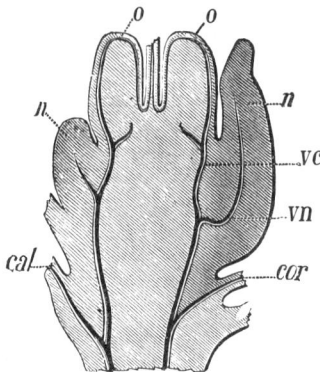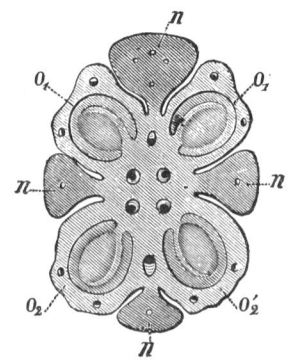

Figure 238. Lengthwise cross-section of the ovary and nectaries of sage: *o*, ovary; *n*, nectaries; *vn*, vessels leading to the nectar glands; *vc*, vessels leading to the carpel; *cal*, calyx; *cor*, corolla.

Figure 239. Cross-section of the ovary and nectaries of the horehound: o_1, o_2, o'_1, o'_2, ovary; *n*, nectaries. (In this and two preceding figures the shaded areas are those where the sugar accumulates.)

Figure 239 shows a similar arrangement (here, cross-cut) in the white horehound (*Marrubium vulgare*), much like that seen in most other Lamiaceae.

We can add another important general observation regarding plants in this family: one shouldn't believe that a given plant is richer in nectar simply because its nectaries may be more prominent or more fully developed. A plant's contribution to honey flow depends above all on how much sugar its nectar contains, and the rate at which another droplet of sugary liquid develops after a bee has collected a previous one.

For example, rosemary (*Rosmarinus officinalis*)—a Lamiaceae that is grown in gardens, and is also quite common in the wild in Southern France—has only slightly developed nectar-producing protuberances, but since they secrete a nectar that is high in sugar and very plentiful, rosemary is an excellent honey plant.

3. *Scrophularia.* The nectar gland of these plants forms an uneven ring around the entire base of the pistil; that ring is covered with small hairs—as is the case with the foxglove, for example. It's interesting to note that bees will even visit a foxglove flower whose corolla

has fallen off, as happens quite often—showing that flowers can attract bees even without brightly colored petals.

4. *Houseleeks (Sempervivum tectorum).* This attractive plant, often grown on rooftops or found on rocks, has red or pink flowers with numerous, radiating petals. There are as many nectaries as petals, and they are arranged in a circle all around the pistil. Figure 240 shows one of these glands, bisected lengthwise. We can see the sugar cells, *n*, which are much smaller than the others and form a highly compact tissue.

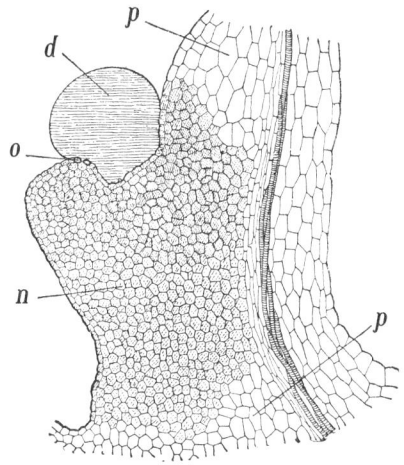

Figure 240. Magnified lengthwise cross-section of a houseleek nectar gland: *n*, nectar gland tissue; *o*, a nectar gland opening; *d*, drop of nectar; *p*, petal tissue (the shaded areas are those where the sugar accumulates).

303. The nectar glands of Scabiosae and Compositae

These plants—whose flowers form tightly packed clusters that appear to be a single large flower (referred to by botanists as a *flower head*)—include many melliferous species. The remarkable thing is that the exterior of their nectaries is poorly developed, while sugar tissues make up a significant volume of the interior, as seen in figure 241 (a partial cross-section of a scabious flower). The segment marked *n* is the only exterior of the nectar gland; the sugar-rich tissues, *s*, are

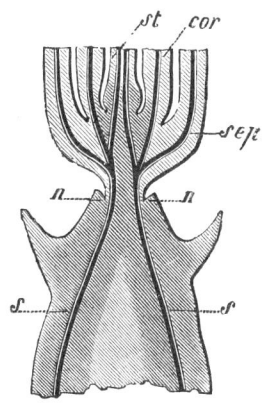

Figure 241. Lengthwise cross-section of the center of a field scabious flower (*Knautia arvensis*): *n*, nectary; *s*, sugar tissues; *st*, stamens; *cor*, corolla; *sep*, calyx (the shaded areas are those where the sugar accumulates).

shaded gray. Things are much the same with the many plants in the Compositae family, whose significant honey plants include cornflowers, thistles, knapweeds, dandelions, etc.

Summary

Nectar glands
Nectar glands or nectaries are the parts of a plant in which sugars accumulate, and which can secrete a sugary liquid, called nectar. They are generally found within the flower, and sometimes on other parts of the plant—for example, on the stipules of the vetch, or at the base of plum leaves.

Sugars contained in nectar and honey
Most nectars contain 70–80% water, a high proportion of sucrose, and a somewhat lower quantity of glucose and fructose.

Honey usually contains 20–25% water, a small percentage of sucrose, and a larger percentage of glucose and fructose.

So honey contains much less water than nectar, and the percentage of sucrose compared to that of glucose and fructose is much lower.

Nectar is transformed into honey, in part, by the bee's digestive tract, which produces a substance that turns a large percentage of the sucrose into glucose and fructose; and, in part, thanks to the evaporation of a large quantity of water before the honey is capped.

Some nectar glands (of the fritillaria, for example) produce large amounts of liquid that is so low in sugar that the bees don't even bother collecting it. Other nectaries, on the other hand (like those of gorse and anemone) exude almost no liquid at all, but they are saturated with a sugary juice that the bees draw out with their proboscis.

Varieties of nectaries
Flowers' nectaries come in all shapes and sizes, depending on the plant, and bees gather their sugary liquid in various ways—from the inside of the flower, or from the side, or through holes already cut through the sepals or petals by bumblebees.

Chapter 22
The Honey Yield Of Various Plants

304. How nectar glands secrete nectar

Generally, the surface of nectaries has numerous stomata—that is, small formations composed of two special cells with a narrow opening in between that communicates between the interior tissue and exterior air (S, S', fig. 242 and o, fig. 240).

When the plant is generating no nectar, these small openings emit nothing but water vapor; but when circumstances are such that large quantities of water are flowing through the plant, this water takes on sugar inside the nectar gland tissues, then emerges in the form of tiny droplets on each stoma, visible when the gland's surface is viewed beneath a microscope (fig. 242) or even with a magnifying glass. If the honey flow persists, these droplets will join together and form a single large drop that

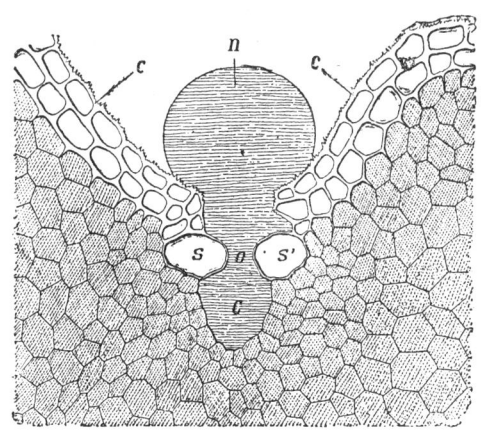

Figure 242. Magnified partial cross-section of a nectary: S, S^1, the two stomata cells through which the nectar, n, is secreted, and accumulates at C; c, sugar cells.

rests on the surface of the nectar gland, or falls and gathers in another part of the flower.

It's very important to note that under circumstances favorable for honey flow, a plant's nectar production cycle is constantly repeated—indeed, if we take a small piece of blotting paper and remove all the nectar produced in a flower, we'll see new droplets forming immediately on the nectary's stomata, then combining into a single larger drop. In the same way, just after a bee has visited a flower, the flower will soon contain just as much nectar as it did before the visit, and another bee won't hesitate to come and collect the latest batch of sugary liquid.

What is more, if a flower is being visited by bees, it will, in the end, have produced more nectar than it would have otherwise. Indeed, if we were to allow a plant to blossom beneath a gauze fine enough to prevent any insect from reaching it, we'd see nectar being produced, but it wouldn't build up indefinitely inside the flower; however, if we withdraw this nectar using a pipette, we'll see that the flower will, in total, produce more nectar than it would have if left untouched.

305. Variation in nectar volume over the course of the day

Nectar production in a given flower can vary highly depending on circumstances.

It varies based on:
1. The time of day.
2. Meteorological conditions.
3. The amount of water in the soil and air.
4. The composition of the soil.
5. The climate.

Let's discuss these factors one by one.

1. *Variation in nectar production at different times of the day.* If good weather is held constant, the volume of nectar in a single flower will drop gradually until around three o'clock in the afternoon, then rise again in the evening—and continue to rise throughout the night, until sunrise.

This general finding has been confirmed by numerous experiments.

The nectar levels of plants protected from insects by a fine gauze have been measured at various times of the day using a graduated pipette.

Such readings were taken every two hours from ten different species of honey plants (stonecrop, lavender, thyme, blue chives, snapdragon, corydalis, wayfarer, phlox, petunia, and fuchsia), over twelve days of consistently good weather. The same variation was observed in each plant, on each day.

For example, the lavender, stonecrop, thyme, and chives showed the following fluctuations in nectar volume:

Nectar volume at various hours of the day, 27 June

Time	Stonecrop (mm^3 per 3 flowers)	Lavender (mm^3 per 10 flowers)	Thyme (mm^3 per 6 flowers)	Chives (mm^3 per 3 flowers)	Temperature in shade		Temperature in sun		Air humid., %
					°F	°C	°F	°C	
5 a.m.	10.0	18.5	1.5	24.0	69	20.5			80
7 a.m.	5.0	18.5	0.5	18.5	73	22.5	75	24.0	74
9 a.m.	1.5	10.0	0.5	5.0	77	25.0	81	27.0	64
11 a.m.	0.5	10.0	0.2	6.0	81	27.0	86	30.0	56
1 p.m.	0.5	5.0	0.05	5.0	82	27.5	89	31.5	55
3 p.m.	0.3	3.0	0.0	3.0	83	28.3	93	34.0	50
5 p.m.	0.2	7.5	0.25	5.0	81	27.0	87	30.5	57
7 p.m.	0.5	10.0	0.5	7.8	75	24.0	81	27.0	70
9 p.m.	1.5	10.0	0.5	8.0	72	22.0			91

As we can see, ten lavender flowers, for example, yielded 18 mm^3 of nectar at 5 o'clock in the morning, 3 mm^3 at 3 in the afternoon, and 10 mm^3 at 9 in the evening. The figures in the table above show that these changes in nectar volume track the changes in air humidity. So the drier the air, the less nectar is found in the flowers.

These results have been confirmed using three other methods, on the same days.

1. A count was kept of the bees returning from foraging over the course of a minute each hour, by forcing them to enter the hives through a long glass-walled passage to facilitate counting. The number of bees entering the hives was found to be greater during late morning and late evening than during the afternoon, which confirms the previous results.

2. Meanwhile, with the hives placed atop precision scales, their

weight was found to be greater in the early afternoon than in the morning or evening. This indicates that fewer bees had left the hive during the afternoon, in keeping with the lower nectar volume in the flowers.

3. Finally, weighing bees (without pollen) returning to the hive at various times on the same day, produced these figures (an average of ten bees):

At 9 a.m.   1.21 g
At 1 p.m.   1.07 g

This result further supports the previous findings, and also shows that each bee returns with a larger harvest when nectar volumes are highest in the flowers.*

In hot weather, the differences we've pointed out become even more noticeable: sometimes, in fact, flowers may only produce nectar in the mornings. On south-facing rocks in the Alps, where houseleek and stonecrop grow abundantly, on a hot day in July, the flowers of these plants may contain significant amounts of nectar during the morning, but not a single drop during the afternoon. So bees actively forage among these flowers in the mornings—but not a single bee is to be seen in the evenings (as observed in Huez, Oisans).

During the summer dry period in Algeria, near the city of Blida, bees are only able to find any nectar to harvest during the early morning. The only time they exit the hive is at the crack of dawn, and they're all back in the hive by eight o'clock that morning.

306. Meteorological variations in nectar yield

Based on the above, we can see that hot, dry weather doesn't favor an abundant and continuous supply of nectar in flowers.

---

*These weigh-ins also reveal that during the morning, when the bees drop ponderously onto the bottom board, they're carrying as much nectar as they possibly can; but that during the afternoon each bee brings less honey than it is capable of, as if the bees feel obligated to return to the hive after a certain period of time.

Generally speaking, the most favorable conditions for a high honey yield in plants are several days of good weather, when flowers are in bloom, following a rainy spell. While the ground is damp, the sunny days that follow encourage plants to draw large amounts of water up through their roots to their flowers, and this movement promotes the emission of sugary liquid. Stormy weather in particular can lead to an abrupt rise in such nectar secretion.

If a rainy period is followed by a long string of consistently good weather, then the favorable effects will build throughout the first several days, then begin to wane as the heat and dryness set in.

This has been shown in a series of experiments conducted both in Louye (in the French department of Eure) and in Paris (in the garden of the École Normale Supérieure) during a string of nice days in June and July. Let's look at one of their findings.

Six fuchsia flowers at the same stage of development, taken each day at 6 a.m. over the course of several consecutive nice days:

	Nectar, mm^3
July 14	250
July 15	340
July 16	450
July 17	180
July 18	160
July 19	105

307. Variations in nectar yield with respect to soil and air humidity

1. *The influence of soil humidity.* All else being equal, when the volume of water absorbed by the roots rises, the volume of nectar emitted by the nectar glands does too.

Let's take a look at the following experiment.

Of two potted blue chive (*Allium nutans*) plants A and B, with equal numbers of flowers, pot A was placed in water, and pot B in low-humidity soil, while the temperature and air humidity for both plants was held constant. Three hours later, nectar volumes were measured in flowers of the same age, resulting in the following averages for three flowers:

Plant A, whose pot was placed in water: 57 mm^3 of nectar.
Plant B, not placed in water: 41 mm^3 of nectar.

Plant A was removed from the water, and two days later the same experiment was conducted—only this time plant B was placed in water, and plant A wasn't. Three hours later, the following averages (for three flowers) were recorded:

Plant B, whose pot was placed in water: 52 mm^3 of nectar.
Plant A, not placed in water: 48.5 mm^3 of nectar.

2. *The influence of air humidity.* All else being equal, as the humidity of the air rises, so does nectar volume in plants.

Let's look at just one of many experiments that have investigated this matter. Two pots of heather, as similar as possible, are kept at the same temperature, in soil with the same level of humidity; the first plant, A, is kept in open air, while the second, B, is kept under a glass cover, with water placed alongside the pot, in air almost fully saturated with moisture.

After twenty-four hours, the following averages were found (for ten flowers):

Plant A, in open air (65% humidity): 18 mm^3 of nectar.
Plant B, under glass cover (98% humidity): 47 mm^3 of nectar.

So the experiment shows that nectar levels are lower when the air is dry.

3. *Artificial means for encouraging nectar production in plants whose nectar levels are naturally low.* If the two conditions mentioned above are in place simultaneously, flowers that don't produce nectar naturally under our climactic conditions can be made to produce it.

So by placing a pot with a non-melliferous plant in water, and keeping the air at high humidity, we may see nectar appear; this experiment has been carried out with flowering plants of common hyacinth, tulip, common rue, *Galium*, lily of the valley, etc., which normally would not produce a single drop of sugary liquid when grown in our regions.

308. Variations in nectar yield due to soil composition

A given plant's nectar yield will vary based on the composition of the soil.

The Laboratory of Plant Biology in Fontainebleau has investigated this matter by conducting experiments on white mustard, buckwheat, sainfoin, alfalfa, wild cabbage, woad, and *Phacelia*.

To ensure that all conditions other than soil type were held constant, square-shaped patches of various soils, each 32" (0.8 m) deep, were placed side by side, but kept separate from each other and the surrounding soil by tiles. The soils included: pure limestone, pure clay, pure sand, and various mixtures of these three soil types.

The plant species listed above were planted simultaneously in these various soils, and two different procedures were used to compare the nectar yield of a given species across the various soils:
1. With the plants covered by large cube-shaped cloth structures to keep out insects, a graduated pipette was used to measure the volume of nectar in flowers of the same age.
2. With the plants left uncovered, the number of bees visiting the flowers within a given time period was counted.

The results were as follows.

The white mustard plant yielded more nectar when planted in soil consisting of limestone and sand, or of limestone, than in soil rich in clay; buckwheat, meanwhile, produced more in silty clay soil than in limestone; *Phacelia* preferred clay soil, or clay and sand; the woad and alfalfa yielded more nectar in limestone; and, finally, the sainfoin's results varied little across the various soils.

309. Variation in nectar yield due to climate

All of the results we've seen thus far make clear that a given plant may be highly melliferous in one region, but not in another. So we can't make absolute statements about whether or not a certain plant is a honey plant. It's more accurate to say that a given plant is melliferous in a given region.

In fact, the nectar production of a given plant varies widely based on latitude and altitude, regardless of the nature of the soil.

1. *Variation by latitude.* Comparative experiments have been conducted in Louye (Eure, France), at 49 degrees latitude, and in Dombås (Norway), at 62 degrees latitude, under essentially identical atmospheric conditions and on plants of the same species (*Silene inflata, Trifolium medium*). These experiments showed that the nectar was always more abundant in Norway than in France. Certain species, like the cinquefoil (*Potentilla tormentilla*) and herb Bennet (*Geum urbanum*) secrete large amounts of nectar in Norway, but are almost completely lacking in sugary liquid in the outskirts of Paris.

In Denmark, bees actively visit various kinds of hawkweed (*Hieracium*), while bees are almost never seen on these same species of plant in France.

So a given plant's nectar yield increases along with the latitude.

2. *Variation by altitude.* If we venture into the mountains of various regions of France, we can find plants and climactic conditions that are quite similar to those at high latitudes—and, as we might expect, a given plant species yields more nectar at higher altitude. This has been confirmed in a precise manner with regard to woad and *Silene*, for example—these two plants produce much higher volumes of nectar at 5,000 ft (1,500 m) of altitude than they do near sea level.

Generally speaking, melliferous vegetation is more abundant in alpine and sub-alpine regions, as indicated by figures for the average harvest per hive, which rises steadily with altitude in the Eastern Pyrenees.*

310. Honeydew and sugary exudations of plants

Generally, *honeydew* is a sugary liquid that falls, in fine drops, from certain trees, covering a tree's lower foliage with spots of varying degrees of viscosity.

Honeydew is produced every year on very hot days—and in certain years, when there is a prolonged dry spell, it can be highly abundant. In this case, it serves as an important resource for bees, although the honey that results is generally lower in quality due to the presence of

---

* See Siau, *Statistique des Pyrénées-Orientales.*

gums and dextrin in the sugary liquid, or, additionally, to the peculiar nature of the sugars that compose it.

Conditions that favor honeydew include, most importantly, warm and dry days, separated by nights that are relatively cool and humid.

Tree species in our region that produce the most honeydew include: oak, ash, linden, maple, poplar, birch, hazel, mountain ash, common barberry, and bramble.

Figure 243. Bees gathering honeydew from oak leaves.

Herbaceous plants such as salsify, scorzonera, many varieties of Brassicaceae, etc., may sometimes produce honeydew as well.

Honeydew may have two very different causes that shouldn't be confused:

1. It is often caused by aphids that attack leaves whose inner tissues are particularly rich in sugary liquid. These aphids only digest a small percentage of the liquid they absorb; most of it is expelled and falls down onto the leaves below in viscous drops—this is called *honeydew* proper.
2. Honeydew can also occur without any aphids, exuded spontaneously through the leaves themselves; in this case, one can see it forming on all stomatic openings, joining together in ever larger drops that finally fall like those we saw above. This is *sugary exudations* of plants.

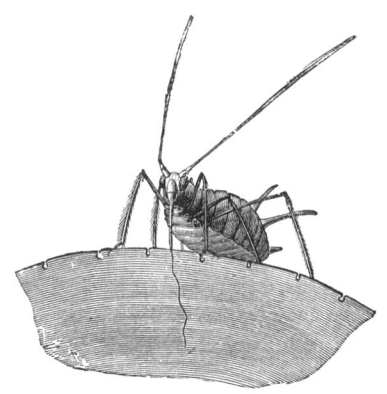

Figure 244. An aphid sucking sugary liquid from leaves (magnified).

You can test the presence of these exudations experimentally by taking a tree branch at the desired time, and placing the leaves of this branch in air saturated with humidity. Check all of the leaves to be sure there are no aphids on them. Some time later, you'll see sugary droplets forming across the entire surface of the leaf, especially on its underside.

This exudation produced spontaneously by the plant is different from the honeydew caused by aphids:

1. It accumulates during the night, and usually disappears during the day, while aphids produce their sugary substance all day and reduce their activity during the night.
2. Its composition is different, being highly similar to that of nectar, while aphid-produced honeydew contains large amounts of dextrin, gums, and, quite often, sugars other than sucrose (mannitol, melezitose, etc.).

Bees only seek out honeydew—especially the honeydew produced by aphids—when there is a lack of abundant honey plants within their reach. This explains why, for example, if the black locust has bloomed abundantly, and, meanwhile, the tree leaves have produced a lot of honeydew, the bees will ignore the latter and visit the locust blossoms exclusively.

The following experiment has shown that bees choose the finest sugary material available to them. Some dishes containing various kinds of plant exudations and honeydew gathered directly from trees were placed near bees' watering trough. Regardless of their origin, the bees preferred exudations from an oak tree to honeydew produced by aphids on a hazel tree. Under other circumstances, they preferred honeydew from linden aphids to the bitter and resinous exudation of the poplar tree.*

CHAPTER 22. HONEY YIELD OF PLANTS

311. How bees allocate foragers among honey plants

A very curious fact—one that is of great interest to beekeepers—is how bees distribute themselves while foraging among various nectar plants.

It seems that on every day of foraging, following the early-morning reconnaissance of the first worker bees to leave the hive, the remaining bees have been precisely informed regarding the location, the relative nectar value, and the distance to each honey plant within a certain radius of the hive.

If you take careful note of the various trajectories the bees embark upon when exiting the hive, and conduct detailed observations of how the bees forage amongst the various plants in the surrounding area, you'll confirm that the workers distribute themselves amongst the flowers in proportion to both the number of plants of a given species, and the value of their nectar. Moreover, as we've just seen in our example of the black locust tree and the honeydew, bees compare, on a daily basis, the quality of the nectars available for gathering on that day.

If, in the spring, for example, after the willows have bloomed, at a time when nothing is in bloom in the fields, and the bees' only remaining resource are the first flowers in the woods, then you can see them actively visiting anemones, lungwort, gorse, and violets. Several days later, the fields of cabbage (and wild cabbage) begin to blossom in large numbers, and one can watch as the bees almost completely abandon the forest flowers—although still in full bloom and rich in nectar—in favor of the cabbage blossoms.

In this way, the bees regulate their distribution amongst the plants on a daily basis, so as to collect the finest sugary liquid available in as short a time as possible.

So one might say that a bee colony, when it comes to both foraging and working inside the hive, manages to distribute its workers in a rational manner by applying the principle of division of labor.

* For more details, see G. Bonnier, "Recherches expérimentales sur la miellée" (*Apiculteur*, 1896). (Sugary exudations of most plants are now believed to be of insect origin. See, for example, *Honey Plants of North America* by John H. Lovell. Ed.)

Summary

How nectar emerges

Nectar is produced on the surface of sugar tissues in the form of small droplets, which usually emerge from the openings of stomata. These droplets gradually join together in ever larger drops; and if one is removed, another eventually forms in its place.

Variation in plants' nectar yield

The quantity of nectar a plant produces varies greatly depending on circumstances.

The volume of nectar drops during the afternoon, and falls progressively over a long series of dry days; the best conditions for nectar production is a string of nice days following a rainy period, or during a period that is stormy but without rain.

The volume of nectar produced rises with latitude and altitude, and with the humidity of soil and air—such that a given plant may be rich in nectar in one region, but not in another.

Nectar volume also varies based on soil composition: a given plant may produce a lot of nectar in a limestone-rich soil, but little in a clay soil—or vice versa.

Honeydew

Under certain circumstances, in early summer, some plants—trees in particular—produce large amounts of a sugary liquid that falls like rain—this is honeydew, which can be an important resource for bees. Honeydew is also produced by aphids, which emit a large amount of the sugary material they consume from the leaves.

Plants also produce exudations that come directly from the leaves themselves, and their composition more closely resembles that of nectar than that of honeydew.

Bee distribution during foraging

At any given time, bees split up to gather the best sugary substances available to them, and do so in proportion to the number of plants of a given species and those plants' honey flow.

Chapter 23

Variations in a Bee Colony's Activity Levels Throughout the Year

312. Bee activity over the course of a single foraging season

Bees' activity levels and foraging patterns during the warm season vary by region.* In Gâtinais, for example, there is little foraging at all, except while sainfoin is in bloom, which lasts around two weeks. This period involves an *intense honey flow*, during which the bees must gather everything they'll need for the entire year, and the beekeeper will harvest the surplus. In a mountainous region, meanwhile, honey plants blossom sequentially, such that during the summer the honey flow remains almost uninterrupted.

The best way to learn how things stand in your particular region is to keep one hive atop a scale. Each day, you can track the hive's weight increases, and at the end of the year, you'll know—with much greater precision than simply by watching the bees—when exactly the honey flow peaks, and when the bees are relatively idle.

When we examine the trends for the foraging season in Fontainebleau, for example, which lasts four months, from mid-May to mid-September, we can distinguish four distinct periods, two of

---

* This chapter is based on experiments conducted by Léon Dufour in Georges de Layens Research Apiary at the Fontainebleau Botanical Laboratory (see *Apiculteur*, 1897 and 1898).

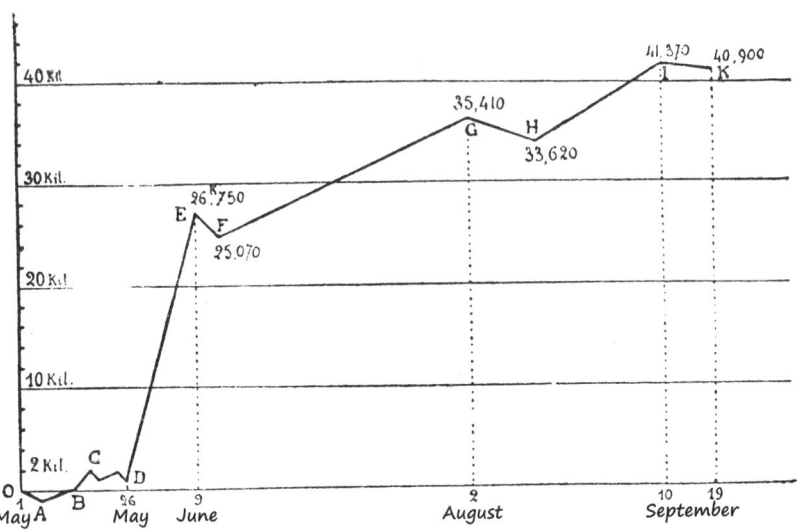

Figure 245. Honey yield of a colony over the course of a season.

which are honey flows. These periods are represented in the graph below (fig. 245); we can describe them as follows.

First period. *Beginning of the season (May). Segment OABCD of the curve.* A loss over the first few days is followed by a rather strong gain, alternating with days of loss. The loss reaches 1.1 lb (0.48 kg), followed by a gain of 5.7 lb (2.57 kg), followed by another loss of 1.9 lb (0.88 kg).

Second period. *Main honey flow (May 26 - June 14). Segment DEF of the curve.* The main spring honey flow is from acacias. The hive population has grown quite large, and the bees are extremely busy. In 7 days, they gather more than half their total harvest for the entire season—45.2 lb (20.5 kg); and in 14 days, they gather two-thirds of the total, 56.4 lb (25.5 kg). On June 1st, which marks the maximum level of foraging, the hive's weight rises by 10 lb (4.5 kg) from morning to evening. The end of the acacias' blooming period is followed by several days of loss (June 9–14) due to the evaporation required for the enormous quantity of nectar that has been gathered.

Third period. *Summer (June 14–August 2). Segment FGH of the curve.* This period is long in duration, and is marked by low honey flow, but a peak hive population, resulting in a decent total yield for the period: 18.5 lb (8.4 kg). The period closes with a stretch of rainy days (August 2–16), causing a slight loss.

Fourth period. *Autumn honey flow* (August 16–September 10). *Segment HIK of the curve.* Common heather, an important honey plant, is in bloom; the bees are extremely active, but the days are shorter and sometimes rainy; the hive population has already fallen off, resulting in a relatively insignificant yield despite the considerable honey flow. The period ends with alternating good and bad days, causing a slight loss for the hive (September 10–18), until finally the loss is sustained day after day—signaling the start of the wintering period.

We can draw up these various periods in the following table:

Period	Duration of the increase, days	Yield		Average daily yield		Loss		Total gain for the period	
		lb	kg	lb	kg	lb	kg	lb	kg
1st	16	5.7	2.6	0.35	0.16	1.1 start 1.9 end	0.5 start 0.9 end	2.7	1.2
2nd	14	56.4	25.6	4.03	1.83	3.7	1.7	52.7	23.9
3rd	49	18.5	8.4	0.38	0.17	3.4	1.6	15.1	6.8
4th	26	13.0	5.9	0.50	0.23	0.8	0.4	12.2	5.5
						Total		82	37.2

If, instead of studying a single hive, we were to track several, we would see that all of them pass through the same stages, beginning and ending at the same time for every hive; the only difference is the work intensity, which varies depending on the strength of each hive. But all the hives, except under truly exceptional circumstances, will go about their work in much the same way. This explains why the information gathered for the single hive kept on the scale is so accurate and useful for the apiary as a whole.

313. Variations in hive weight over the course of a day

If the yield on a given day is low, then the weight of the nectar that is brought back now and then to the hive is of little significance; if there is much change in a hive's weight under such circumstances, it's because the number of bees away foraging varies at different times of the day.

COLONY'S ACTIVITY DURING A DAY

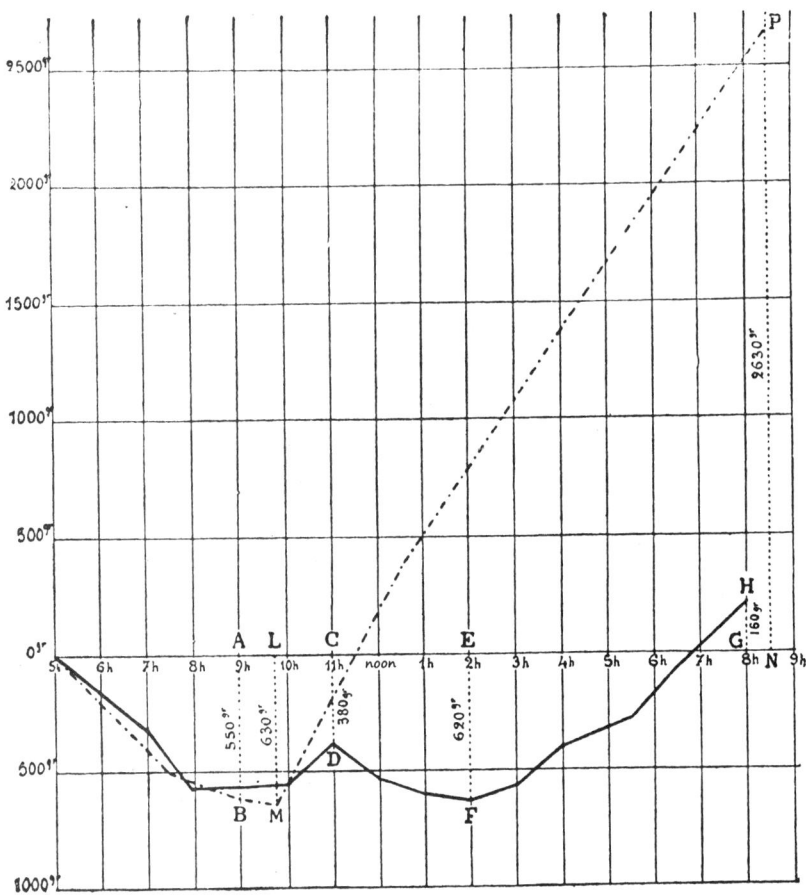

Figure 246. Honey yield of a colony over the course of a day. The dotted line indicates a day of intense honey flow, while the solid line shows a day of low honey flow.

The solid curve on the graph in figure 246 shows the fluctuations in weight over the course of a day of good weather. On this particular day, at 5:30 a.m. all of the bees are still in the hive; as they gradually leave, the hive's weight drops. But around midday the flowers contain less nectar than in the morning or evening, and more bees return than leave; so, after an initial drop, the hive's weight rises again. Later, the nectar becomes more plentiful, many bees leave the hive, and the hive's weight falls again. Finally, at the end of the day, the bees gradually return to the hive, whose weight rises steadily until evening—and once all of the bees have returned, the registered increase in weight for the day represents the day's total yield.

We can measure the maximum points of loss, *AB* and *EF*, separated by the minimum *CD*; at the end of the day, the overall gain for the day, *GH*, is slight.

However, during heavy honey flow (see the dotted curve in figure 246), the weight of the collected nectar soon makes up for the weight of the departed bees, and the pattern is different. As the day begins, the hive's weight falls as more and more bees depart, but soon it rebounds, and continues to rise right up until evening, when the scale reveals a considerable weight increase due to the large quantity of nectar collected. The maximum weight loss, marked *LM*, comes quite early in the day, and the gain at the end of evening, *NP*, is considerable.

314. The yield is not always proportional to the bees' activity level

If there is time for it, a beekeeper should take frequent strolls through the apiary in order to keep track of everything as it unfolds, and form an idea of the various hives' activity levels at various times of the year. But observations gathered in this way can lead to misconceptions.

On some days, the bees will appear highly active, entering and exiting the hives in great numbers. Yet, come evening, you may be surprised to find that the total yield for the day has been low. Why?

Because on that particular day the flowers were extremely low on nectar. When this is the case, the bees don't wait until they are loaded down with as much nectar as possible before returning to the hive—instead, they carry only a small amount on each foraging run. This explains why the day's yield is lackluster despite all of the coming-and-going. So while a superficial glance around the apiary might lead us to believe that the yield has been high, the scales provide us with more accurate data.

315. Honey consumed during winter

In the region surrounding Fontainebleau, you should harvest honey and winterize your hives around mid-September.

WINTER HONEY CONSUMPTION

Up until mid- to late October, the weather is sometimes good, and the bees remain quite active. So, on occasion, there may still be a modest yield on a given day—but in general, the bees consume a great deal, eating honey to maintain their high activity level. For three hives kept on scales, between 3.3 and 5.5 lb (1.5–2.5 kg) of honey was consumed during this period.

Then, over November, December, January and part of February, the cold season sets in, and the bees remain clustered inside the hive, eating little, aside from those rare days when intense sunshine may lure them outside. During this period, each hive consumed between 11 and 13 lb (5–6 kg) of honey. Finally, from mid-February to May, the bees gradually resume their activities, the queen resumes her egg-laying, and a great amount of honey is used to feed the larvae; the total amount consumed by each hive during this period ranged from 17 to 20 lb (7.5–9 kg).

So the total amounts consumed by each hive during the winter came to: 33 lb, 38 lb, and 32 lb (15.2 kg, 17.2 kg, and 14.7 kg). Keeping in mind that the winter of 1896–97 was quite mild—and that considerable amounts of honey was consumed because spring got off to a rainy start—we can conclude that these amounts consumed were quite high, and that, generally speaking, one should leave 33 lb (15 kg) of honey in a hive, to avoid having to inspect the hives too early in the season to check whether any require feeding.

Moreover, if you have moveable-frame hives, it's wise to always keep a few honey frames on hand, with the honey still intact, until the following season. If for some reason certain hives need a feeding boost in the spring, you can give them a frame to tide them over until the warm weather sets in, without all of the inconveniences of feeding by other methods.

Chapter 24
Managing an Outyard

When you've grown from a beginner into *true beekeeper*, and wish to increase your number of hives, you'll need to keep multiple beeyards, since having too many colonies in a single location will inevitably ruin the apiary during bad years, unless you go to great lengths to feed your bees during autumn. If, meanwhile, the same number of hives are distributed among five or six yards, at a distance of several miles from each other, then the bees—now with fewer of them at any single location—will manage to produce at least enough honey for their winter reserves.

Now I'll attempt to describe in detail all of the procedures required throughout the course of the season to manage an outyard, spending as little time doing so as possible. As an example, I'll take my own apiary, located in the forest of Beaupuits, which is around 2 miles (3 km) from my home.

General considerations

When you're first planning to set up an apiary, it's extremely difficult to determine the area's potential honey yield without simply giving it a try. Beekeepers who see great results from their apiaries typically owe their success more to the local terrain, climate and flora than to any other factor.

For a long time now, in my own apiary and in those of my neighbors, I've been transferring a large number of fixed-comb hives into moveable-frame hives. But, come autumn, these hives have failed to produce more than a few pounds of surplus honey, while in Savoie—a region extremely rich in nectar—hives that were transferred the year before have often produced more than 50 lb of surplus honey the next fall. Here's another example: the linden tree, which in some regions provides a great deal of nectar, hardly produces any in my area. In my apiary, I've never seen the hives I keep atop scales gain much weight while the lindens are in bloom.

Long experience has taught me that one should never stray from the following three principles, which constitute the bedrock of any good beekeeping:

1. Large hives.
2. Large frames.
3. Enough room in the hive to always allow the queen's egg-laying to proceed uninterrupted.

In this way, you'll: 1) minimize natural swarming; 2) produce as much honey as the location will allow; 3) build up a strong population in the fall. All three points are of great importance for the future of the apiary.

My neighbor farmers have long applied the simple method to their apiaries, which consist of horizontal hives. These procedures have been used in my apiary for twenty years now, and for fifteen years in theirs. I should add that this simple method is becoming increasingly popular in various parts of France.

The apiary

My hives are in the woods, about 2 miles (3 km) from my home. They're arranged in two rows, down a path shaded by large trees. The hives are placed in pairs atop stands, 16" (40 cm) off the ground. This allows me to handle them much more conveniently, without stooping—and, when they're kept higher off the ground, they're less susceptible to excess humidity during the winter.

The apiary is surrounded by wooden pickets connected by four strands of barbed wire—the most economical form of fencing. There's a stream near the apiary that always has water in it, and the bees need only fly a hundred yards (100 m) or so to reach the meadow.

Based on my own observations, in comparing various apiaries, I've concluded that the bees in an apiary shaded in a forest tend to swarm less often than one kept in full sunlight.

For several years now, based on observing other beekeepers, I've realized how important it is not to arrange the hives in a straight line, but rather as irregularly as possible. This arrangement keeps the bees from returning to the wrong hive as often after foraging, and allows the queens to find their own hive more easily when they return from their mating flight, thus leading to fewer orphaned hives.

The equipment

My equipment is extremely simple, and can all be transported in a wagon.

1. A wheelbarrow with a frame box placed on top, large enough to hold twenty frames or so. This wheelbarrow and frame box are used to hold any frames I've just removed from a hive, or am planning to put there;
2. Two frame boxes large enough to hold 50 frames each, for moving full or empty frames from the apiary to the house or vice versa;
3. A good smoker;
4. A bucket with a tight-closing lid, and a long kitchen knife for trimming and removing bits of comb when repairing irregularly built comb;
5. A chisel (hive tool) for easily detaching frames. To use it, simply slide it in between two frames, then use it as a lever to detach any frame, even one stuck fast with a lot of propolis;
6. A small broom made of several goose feathers, for sweeping bees off comb.

Notes on inspecting hives

You should get used to visiting your hives without regard to what mood the bees are in, since waiting for an ideal day each time you need to inspect would waste too much precious time. The bees may be quite easy to work with one day, and more difficult the next. In any case, with a bit of experience and a lot of smoke, you can always get the job done. I've managed to harvest honey in October with neither me nor my assistant being stung a single time.

When opening a hive, the most important thing is not to remove a single frame until you've smoked heavily for a bit, from the top, in between the combs. Then, you can move quickly.

When I'm driving the bees off the frames, I never brush them outside the hive, onto the bottom board, but always back into the hive. In this case, they fall to the bottom of the hive; a few will fly away, but they'll soon return to rejoin the other bees.

While I'm working, I always use an assistant, who smokes the bees while I'm busy; I've found that two people working in this way can easily do the work of three or four people working separately.

When winter ends, novice beekeepers are often in too much of a hurry to inspect their colonies. This is a mistake—you should never disturb your bees while they're at rest. Instead, wait until they've been actively working for a week or so. If you inspect your hives too soon after a long, harsh winter, some colonies may be found to have no brood, and a novice beekeeper may assume that they've become queenless, when in fact the queen simply hasn't begun laying eggs yet. However, if you haven't left enough honey for wintering (40–44 lb, or 18–20 kg), some colonies may be running low on provisions, and you'll have to feed them.

Here's the easiest way to feed a colony in spring. Using a saw, cut several slices from a sugar loaf. Uncover some frames and, after soaking a sugar slice slightly in water, place it on top of the frames, then cover it with some thick fabric to prevent heat loss.

There's another very simple way to feed bees: dissolve 17 lb of sugar, while heating, in 1 gal of water (10 kg sugar, 5 L water). Once the resulting syrup has cooled down, use a thin-spouted burette to pour it, from a certain height, into the cells of some well-built comb, in a frame laid flat on a table. Then, cover the frame with a sheet of paper, turn it over, and fill the other side as well.

It's not a bad thing if a bit of old honey remains in the hives when main honey flow begins; when this old honey is put through the extractor along with the new, it will contribute greatly to good, quick crystallization.

General inspection of the hives

Before going further, I should mention that my horizontal hives are arranged quite differently now than they were at first. Each hive has two entrances, one on each end, and the brood nest is always at one end of the hive, instead of in the middle—either to the left or right, facing one of the two entrances, *with the other entrance always kept shut*. Since I believe that division boards are of no use, I haven't used them in my hives for a long time now. But I have saved a few of them to use when transferring a colony or installing a swarm in a hive, in order to force the bees to build exclusively within the space I've given them.

I position my wheelbarrow, carrying the box full of frames, behind each hive, and arrange all of them as follows.

Let's take, for example, a colony that wintered on 18 frames (fig. 247). I begin my inspection with frame A, the one closest to the hive wall, near the entrance. I remove it, then continue down the row until I come across the first frame with brood—in this case, frame D. So at this point, I've removed the first three frames (A, B, C), which contain various amounts of honey, and placed them, along with their bees, at the other end of the hive (fig. 248); these three frames will be replaced by three dry-comb frames with worker cells (U, T, S—to the right in fig. 248), which I'll take from the hive itself or from my box of reserve frames. If you don't have any extra frames available, you can use frames with foundation instead.

But before putting the empty frames U, T, and S in place, I check the condition of the brood in frames D, E, F, G, and H (fig. 247). I leave these frames where they are, then remove, for example, frames I and J, which contain no brood but a lot of honey, and move them up against frames A, B, and C (fig. 248), and, in their place, I insert two worker cell frames, P and R (fig. 248). The remaining frames—K, L, M, N, and O—are left where they are, and I add more new frames until the hive is entirely filled.

So when the procedure is complete, the hive looks like this (fig. 248):

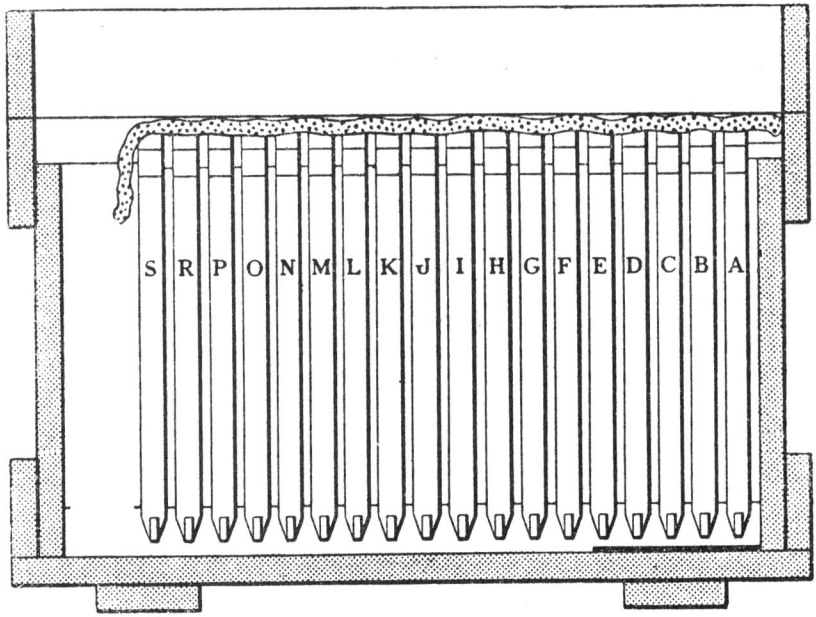

Figure 247. A hive prepared for winter.

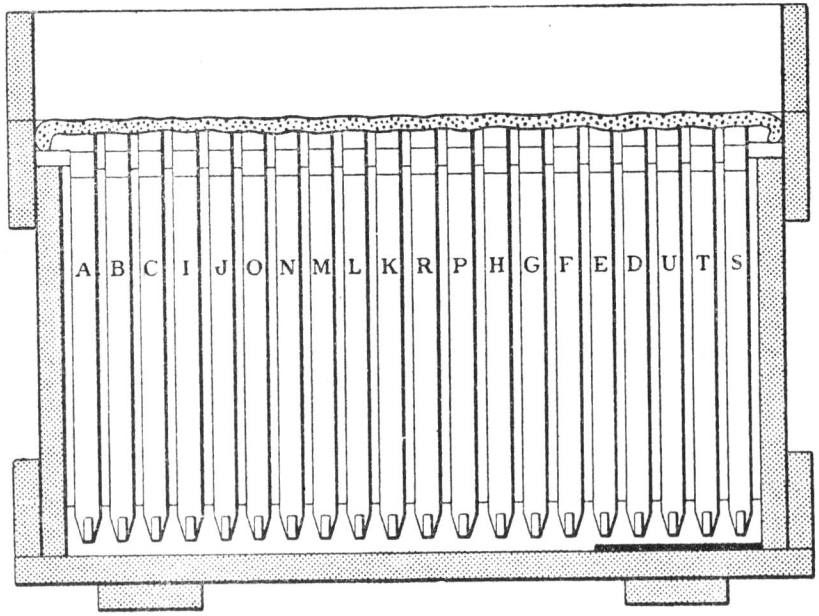

Figure 248. The same hive as in fig. 247, after spring inspection.

Three empty frames of drawn comb	S, T, U
Five frames of brood and honey	D, E, F, G, H
Two drawn-comb frames, empty or with some honey	P, R
Five empty frames of drawn comb	K, L, M, N, O
Five frames containing various amounts of honey	J, I, C, B, A

Of course, all of these instructions aren't absolute in terms of the number and condition of the frames that are moved—this all depends on the size of the colony and the amount of honey found in the hive. However, the general order the frames should be put in is always the same, and can be summed up as follows.

In spring, the brood frames should be surrounded by frames of drawn comb that are either empty or containing only little honey, so that the queen will not run out of room for laying eggs, right up until the period of the main honey flow. Meanwhile, the frames containing a lot of honey should be moved to the far end of the hive—the end furthest from the brood.*

I arrange all of my frames in this manner each spring, giving the queen plenty of room to lay. Most of the old honey, if any remains in the hive by the time the main honey flow arrives, will be put through the extractor along with the new honey.

Between the first inspection and the honey harvest, all that's necessary is an occasional stroll through the apiary to make sure the bees are working normally.

Robbing is no threat to remote outyards, as long as you're keeping local bee races. The only case when you'll need to keep an eye on your colonies is when you're forced to feed them—and this should never happen, assuming you leave them enough honey in the fall.

If, during one of these external inspections, you find a colony whose bees are exiting the hive with pollen on their legs, and bustling

---

* You should use the spring inspection as an opportunity to add some frames to the hives—frames primed simply with some old pieces of comb attached with hide glue to the top of the frames. These frames should be placed at the far end of the hive opposite the one where the brood is found, and should be interspersed with frames already built out with comb. Add two or three such frames to each hive. This will ensure that the bees build the comb regularly within the frames.

around the bottom board as if worried—and, at the same time, the bees seem to be working sluggishly or irregularly—then you can be almost certain that the colony has become queenless, or has a drone-laying queen. If this is the case, give the colony a frame with brood of all ages, or a queen, or a swarm.

In the spring, it's very important to check the brood's condition carefully. If the brood is in tightly arranged patches, then the colony is in good shape: its queen is probably fertile, and its future is bright. This holds true regardless of how strong or weak the colony seems, or how much brood it contains.

If, however, you find that the brood is scattered, then the queen can't be doing well. In most cases, the hive will replace its queen and eventually be in good shape for wintering and for the following season.

You may also discover that a colony is orphaned, or has a drone-laying queen. If this is the case—and it does happen from time to time (if not every year, in a large apiary)—then you can eliminate the now worthless colony using the simple method described below. But save this procedure for a nice day when the bees are highly active, which clearly indicates that there is nectar in the flowers.

Move the hive to the edge of the apiary, remove all of the frames, and brush the bees off onto a bottom board placed on the ground, in the sunlight. The bees will fly away, and, unable to find their old hive, will seek out the hospitality of the colonies surrounding their former home. The honey this orphaned hive contains can go toward the purchase of a replacement colony.

If you're not able to carry out this procedure on the same day, you should greatly narrow the hive's entrance for the time being to avoid robbing.

The best way to replace queenless hives is to set aside funds from honey sales for buying replacement colonies, since, by buying your colonies from other keepers, you'll be constantly introducing fresh blood into your apiary. However, if you have multiple outyards at some distance from one another, you can also keep a small apiary of colonies in fixed-comb hives near your home, or moveable-frame hives with only a small number of frames inside—this apiary will be used to produce swarms for you to collect. This breeding apiary will allow you to fill in the gaps in your outyards as they arise.

The honey harvest

To obtain honey that is of high quality and has a long shelf life, you should only extract it from the combs when they are almost entirely capped. This means that I harvest late in the year, in September. Of course, this means that my sainfoin honey is mixed with honey from fall flowers—but since, in order to find buyers easily, I have to sell my honey at roughly the same price set by local honey suppliers, and since the honey I produce using an extractor is infinitely superior to theirs, this medley of honey from various flowers presents no real problem. In any case, it's convenient to harvest in the fall, since by this time of year there are fewer bees in the colonies, which makes the entire process go more quickly.

When harvest time comes, I make my way to the apiary with all of the necessary equipment, just as in spring. I harvest at around four or five in the afternoon, and finish up around dusk. My method allows me to harvest a dozen or so hives in two hours.

I open each hive and remove the frames one by one, starting with the one that is furthest from the brood nest. If this first frame contains a lot of honey, it's a sure sign that there's a lot of honey in the hive. I usually remove between 8 and 10 frames from a hive that contains 20 frames and is heavy. Note, however, that even if the colony possesses less honey, I'll still need to remove the same number of frames—the only difference is that only a third or a quarter of these frames' surface will be taken up by honey, instead of the entire surface. Once you have a bit of experience, I recommend lifting the hives from the back and, by comparing their weight, deciding whether to remove a frame or two more or less. In any event, I leave at least 38 lb (17–18 kg) of honey for wintering.

Once the surplus frames have been removed, I typically cover gaps between the remaining frames with metal strips folded into a V shape (or with small wooden slats), then cover the empty portion of the hive with more slats, being careful to leave a 3/16" (0.5 cm) gap between them to allow the vapors emitted by the bees during the winter to escape. Finally, I spread a wool blanket or some straw on top of the frames and slats.

Wintering

As you probably noticed, I prepare my hives for wintering simultaneously with the harvesting procedure, which allows me to avoid returning to the apiary until the following spring.

On the day when I finish these procedures, I place a mouse guard in front of each hive entrance located on the same side as the frames; the other entrance is always kept shut. As for the frames I've run through the extractor, I don't leave them in this same apiary for the bees to clean; instead, I place them in another apiary near the home of a tenant on my property.

But if you don't have enough space to store the frames during the winter, you can put most of them back into the hives. For example, you can leave 18 of them in a 20-frame hive, covering the last frame with wool blanket or straw as shown in fig. 247; the extra space allows any humidity—so harmful to bees—to escape through the roof during the winter.

Sometimes I'm forced to harvest my honey earlier in the year, at a time when the honey has been only partially capped. In this case, I pass the frames through the extractor without uncapping them, to extract the uncapped honey only, which I use to make mead. Then I uncap the capped cells and extract the honey from them as usual.

I've tried just about every kind of knife for uncapping cells, but the one that has the most advantages is one that looks a lot like a drawknife used in woodworking: a curved blade, 3/8" (1 cm) shorter than the length of frames, with a handle on each end, allowing it to uncap easily, and not too tiringly, since it can be held with two hands. The knife should be heated before use.

All in all, managing an apiary throughout the entire beekeeping season can be reduced to the following two main procedures: 1) the spring inspection, when the hives are filled out with frames; and 2) the fall harvest and preparing the colonies for wintering.

In these few pages, I hope I've shown how you can achieve, as simply as possible, a profitable average return over a number of years which, individually, may be more or less favorable—and how you can manage multiple apiaries in little time and with low risk.

As I said at the outset, the advice I've given here is not meant for those who are just beginning to keep bees—but an intelligent beginner won't take long to learn how to handle hives and the bees in them. And it's beginners like these—novices who have grown into true beekeepers—that I've had in mind while writing this chapter.

How many farmers, priests, teachers and even industrialists would like nothing more than to manage bees on a fairly large scale, but are prevented from doing so by their many other responsibilities! Many of them may even have a few colonies near their homes, but balk at the idea of establishing larger apiaries when they consider all of the complex procedures that would constantly occupy them throughout the bee season—not to mention the daily monitoring that beekeeping manuals assure them is indispensable.

If you want to have a productive apiary, without it pulling you away from your other pursuits, then the advice I've given will allow you to increase your revenue prudently, steadily, and almost risk-free, and allow France to reap the harvest of honey from its abundant nectar plants—a precious resource that would otherwise go to waste.

Index

NUMBERS REFER TO SECTIONS

A
Abdomen of a worker bee 9
Acacia 50, 52, 53, 300
Acherontia atropos 292
Adair hive 213.1
Adding supers
    for the fall honey flow 186
    under fixed-comb hives 120
    when, to vertical hives 179 & on
Advantages of
    foundation frames 119
    movable-frame hives 47
    vertical hives 173, 174
    wax foundation 48
Aesculin 299
Aesculus hippocastanum 52, 299
Afterswarm, *see* Swarm
Afterswarm prevention 233
Agriculture, importance of beekeeping for 4
Ajuga 50
Alfalfa 51
Allowing bees to draw comb, advantages 247
Alpine rock-cress 51
Anemone 295, 301
Antennas of a worker bee 9
Anther of a stamen 17
Anthophora 14

Ants 292
Aphids 310
Apple tree 17, 52
Apprenticeship of a beginner 77
Arabis alpina 51
Artificial swarming, *see* Swarming
Arugula 301
Asclepias 293
Asters 51
Atomizer 241

B
Badger 292
Barometer 219
Baskets (pollen) of a worker bee 9
Basswood 50, 52, 53
Bead of wax, priming frames with 102
Beans 51, 300
Bearding 12
Bee beetle 292
Bee bread, *see* Pollen
Bee brush 222
Bee escape 226
Bee louse 292
Bee repellents 223
Bee(s)
    at the hive entrance 5
    Carniolan 242
    collecting pollen 17

collecting propolis 18
collecting water 19
feeding on syrup 90
gathering outside the hive 12
insects that can be confused with 14
inside the hive 20
irritable 64
Italian 242
lifespan of 32
making come-hither signal 104
male 8, 10
old 31
on flowers 13, 15, 311
pacified 58
queen 21
races, foreign 242
work performed during her life 31
worker, development of 36
young 31
Beehouse 218
Beehouse for German hives 214
Beekeeper
  amateur 252
  hobby 252
  professional 251
  sideline 250
Beekeepers, different kinds of 249
Beekeeping
  its products 253
  regions more or less favorable for 53
  the future of 2
Beeswax, *see* Wax
Beewolf 292
Beginners
  and movable-frame hives 56
  apprenticeship 77
Bismuth subnitrate, mead making 262
Black locust 50, 52, 53, 300
Black medic 51, 53, 300
Blackberry 50
Blake hive 212
Block of wood for installing foundation into sections 192
Blueweed 50

Bombus pascuorum 15
Bombus terrestris 15
Bottom board 72
Boxes with sections 175, 191
Brandy, honey 275
Brassica 15, 301
Braula (bee louse) 292
Bristlegrass 293
Brittany honey 257.1
Brood 26, 27
  compact 137
  development of 33
  drone 27
  drone in worker cells 35
  quality 137
  ring-shaped 137
  scattered 137
  worker 26
Broom plant 17
Brunet hive 212
Brush 222
Buckwheat 50, 51, 301
Bugleweed 50
Building movable-frame hives inexpensively 58, footnote
Bumblebee, large earth 15
Bumblebees 14, 15
Buying
  colonies 65, 66
  stocked movable-frame hives 229
  swarms 69

C

Cabbage 15
Calluna vulgaris 50
Cap 44
  hives 44
  managing cap hives 207
  movable-frame 208
Capped honey 24
Cappings 24
Carder bee 15
Carniolan bees 242
Castor-bean 298

Cells 20, 22, 23
  built at an angle 23
  drone 27
  queen 28, 37
  with honey 24
  with pollen 25
  worker 23
Centaurea jacea 50
Chalicodoma 14
Chalkbrood 289
Chamonix honey 257.1
Cherry 52, 298
Cleaners 6
Climate
  favorable for beekeeping 54
  Mediterranean 54
  mountain 54
  nectar resources variation with 309
Clover
  alsike 51
  crimson 50
  red 300
  white 15, 51, 300
Collapse of combs 155, 156
Collecting
  swarm 104
  afterswarm 111
Collection of pollen by bees 17
Colonies
  buying 65–68
  uniting 235
Colony 29, chapter 2 summary, *see also* Hive
Columbine 299
Comb(s) 20, 22
  collapsed 155, 156
  construction by bees 29
  honey, reserve 168
  honey, weight of 124
  inspecting 169
  movable 46
  new and old 30
  spacing 20
  treating with sulfur 86
  unused during transfer 149
  uses of 101
Combs (brushes) of a worker bee 9
Come-hither signal 104
Comparison of methods 245 and chapter 28 summary
Composition of
  honey 257.2
  mead 269
  nectar 295–297
Containers for honey 129
Cornflower 51, 303
Cyser 273

D

Dadant hive 172
Dandelion 51
De Beauvoys hive 212
Death's-head hawkmoth 292
Derosne hive 212
Description
  horizontal hive with movable frames 98
  vertical hive with movable frames 172
Devauchelle hive 213.2
Development
  drone 38
  queen 37
  worker 36
Diseases of honey bees 283
Dissolving a hive 85
Division board 227
Division of labor among bees 31
Dragonflies 292
Drawing comb, bees 29
Driving bees from one hive into another 146
Drone fly 14
Drone trap 224
Drone(s) 8, 10
  brood 27
  brood in worker cells 35
  cells 27

development 38
evicted by bees 122
number of in a hive 31
trap 224
Drone-laying queen 35
hive with 84
Drumming 146
Dysentery 288

E

Echium 50
Egg-laying by the queen 33, 34
Eggs 36
finding on black sheet 146
number the queen can lay per day 34
Embedder for wax foundation 98
Enemies of the bees 283
English hives 213.1
Entrance bees at 5
Equipment, section honey 175, 191
Equipment, vertical hive 171
Erica cinerea 50
Eristalis tenax 14
Eruca sativa 301
Eucera 14
Evicting drones 122
Experiments on the quantity of honey required to produce wax 161
Extracting heather honey 167
Extracting honey from movable-frame hives 130
Extractor 47, 130, 225
budget 225
reversible 225
Eyes of a worker bee 9

F

Fanning bees 6
number of 96
Feeder
Derosne 220
English 220
Layens 220

Feeding
fixed-comb hives 87, 89–91, 205
movable-frame hives 126, 127
movable-frame hives in the spring 140
problems 141
speculative 231
stimulative 231
sugar paste 232
swarms in case of bad weather 109
Fermentation of honey 265
Field mice 76, 292
Filament of a stamen 17
Filtering honey 129
Finding queen 237
Fixing queenless hives 236
Flipped-hive transfer 143
Flowers
bees on 13
pierced by bumblebees 15
visited by bees 15
Follower board 227
Foraging season 95
end of 122
Foraging, judging how it is going 96
Forget-me-not 51
Fork, uncapping 222
Foulbrood 283–285
treating 286, 287
Foundation 48
advantages 48, 119
how to install 99
Foxglove 302
Frame box 117
Frame grip 118
Frame(s)
arrangement in the spring 161
brood frame of a vertical hive 172
for a super of a vertical hive 172
holding sections 194
movable, hives 46
prepared for transferring 145
primed 48, 100
sections 175

sizes and shapes 210
wired for wax foundation 99
with strings for direct transfer 145
Future of beekeeping 2

G
Galleria 290, 291
Gâtinais honey 257.1
German hives 214
Germander 50
Glastum 50, 130
Gloves 57
Glucometer, Guyot 263
Glucose 295
Goldenrod 50
Gorse 295
Grafting queen cells 239
Gravenhorst hive 212
Green beans 15, 51, 300
Guard bees 6
Guyot glucometer 263

H
Harvest of honey by beekeeper 123
Harvesting
   fixed-comb hives 42, 201, 244
   frames from a movable-frame
     hive 125
   honey, tools for 129
   supers 188
Hawthorn 298
Head of worker bee 9
   with pollen plumes 289
Heather 50, 52, 301
   bell 50
   common 50
   honey 167, 257.1
   nectar 297
Hellebore 299
Helleborus 299
Hive 42
   album 212
   American standard 213.1
   Arabic 215

bottom board 72
buying 65–68
cap 44
cap, managing 207
casting a primary swarm 106
casting an afterswarm 112
cork oak bark 42
Corsican 215
Dadant 172
dead 83, 85
disorganized/queenless 84, 85
English 213.1
feeding required 88
fixed-comb 42
   supers 45
   fall feeding 205
   feeding 87–91
   harvesting 42, 201, 244
   inspecting 79
   managing 195
   monitoring 120
   smoking 79
   transferred to movable-frame 142
   uniting after harvest 203, 204
   wintering 76
French 98
German 214
gum 42
horizontal 98, 212
in excellent shape after winter 19
Langstroth 172
Layens 98
leaf, Huber 212
location 71
log 42
long 212
movable-comb 46, 47
movable-frame 46
   advantages of 47
   and beginners 56
   description 98
   different systems 210, 211
   disorganized 128
   empty, practicing 116

INDEX

feeding 126, 127, 140
installing a swarm 107
monitoring 121
populated, buying 229
preparing for installing swarms 97
selling 118
smoking 118
uniting 132
vertical 171
wintering 134
observation 20, 217
parent hive (artificial swarm) 163, 200
price of 70
probing 88
purchased in another region 164
purchased, value of 67
queenless 84, 131
Quinby 172
relocated (artificial swarming) 163, 200
Scotch 215
skep 42
stacked 45
stand 72
straw 42
strong, wintered poorly 81
transferred, monitoring 151
transporting 74
two-queen 213.2
uniting 235
vertical with supers 171
  advantages 173, 174
  different 213.1
  equipment 171
  management 176
warm-way 214
weak, how to strengthen 154
weak, wintered well 80
weight 67, 68, 88, 96, 312–315
wicker 42
with a drone-laying queen 85
without honey 82
wrapped for transport 74

Hive entrance bees at 5
Hive stand 72
Honey
  brandy 275
  Brittany 257.1
  capped 24
  Chamonix 257.1
  comb 174, 191
  comb, no sections 243
  composition 257.2
  containers 129
  dark amber 257.1
  filtering 129
  Gâtinais 51, 257.1
  harvesting 123
  heather 167, 257.1
  jars 129
  Landes 257.1
  processing 202
  production by plants 304–309
  red 257.1
  sainfoin 257.1, 297
  sections 174, 175, 191, 192, 194
  sections for horizontal hives 194
  selling 256
  settling tank 129
  storage 255
  to be left for the winter 125
  tools for harvesting 129
  uncapped 24
  uses 276
  varieties 257.1
  vinegar 274
  virgin 202
Honey comb reserves 168
Honey plants 50–52
  bees visiting 311
Honey room 254
Honey yield of plants 304–309
Honeydew 16, 310
Honey flow 95
  end of 122
  judging the amount 96
Hop clover 51

Horehound, white 302
Horizontal hives, varieties of 98, 212
Hornets 292
Horse chestnut 52, 299
Houseleek 302
Huber leaf hive 212
Humming of pacified bees 58
Hydrometer 267
Hygrometer 219

I
Importance of beekeeping for agriculture 4
Increasing the number of hives 162
Influence of climate on nectar resources 54
Influence of soil 55
Insects that can be confused with bees 14
Inserting foundation into sections 192
Inspecting
   fixed-comb hives 79
   hives at the end of the season 124
   hives early spring in year two 136
   movable-frame hives 118
   supered hives 187
   vertical hives after harvest 189
   vertical hives in the spring 177
Installing foundation 99
Installing swarms into frame hives 95
Introducing a queen into a hive 240–242
Irritating bees 64
Isatis tinctoria 50
Italian bees 242

J
Jars for honey 129
Journal 166

K
Knapweed 50, 303
   brown 50

L
Labor, division of 31
Lamiaceae 50
Landes honey 257.1
Langstroth hive 172
Larvae, bee 36
   worker 26
Lathyrus 51
Lavandula 54
Lavender 54
Layens
   feeder 220
   hive 98
   smoker 61
Leadwort 298
Legs, worker 9
Lifespan
   queen 32
   worker 32
Linden 50, 52, 53
Liquidating a hive 85
Lizards 292
Location of hives 71
Lombard hive 215

M
Maintaining the number of hives 162
Managing, *see also* Procedures
   apiary chapter 12 summary, chapter 24
   cap hives 207
   fixed-comb hives 195
   vertical hives 176
Mandibles of worker bee 9
Maple 50, 52
Marrubium vulgare 302
Mason bees 14
Materials gathered by bees chapter 1 summary
May sickness 292
Mead 258
   alcohol content 260
   composition 269
   fining and bottling 266

how to make 262
  sweet 268
Mead, grape 270, 271
Medicago lupulina 51, 53, 300
Medicago sativa 51
Mediterranean climate 54
Megachile 14
Melilotus 50
Melliferous plants 50
  trees 52
Meloe beetle 292
Melomel 270, 271
Merchants, honey 66
Methods, compared 245 and chapter 18 summary
Mice 292
Microscope 219
Milkweed 293
Mint 50
Mites 292
Model apiaries 3
Monitoring
  fixed-comb hives 120
  hives during season 153, 165
  movable-frame hives 121
  supers 183
  transferred hive 151
Mother, *see* Queen
Mountain climate 54
Munn hive 212

N
Narcotism 289
Nasturtium 299
Nectar 15, 294
  collected outside flowers 16
  composition 295–297
  secretion 304
Nectar plants 50
Nectaries (nectar glands) 294, 304
  of different plants 298–303
Nucleus hives 239
Number of
  drones in a hive 31
  eggs the queen can lay per day 34
  fanning bees 96
  hives (maintaining and increasing) 162
  workers in a hive 31
Nymph (pupa) of a honey bee 36

O
Observation hive 217
Oil can, water-bath 221
Old bees 31
Onions, flowering 51
Onobrychis 51
Operations, *see* Procedures
Order of frames in the spring 139, 161
Orientation flight, young bees 11
Orphaned hive, *see* Queenless hive
Osmia 14

P
Pacified bees 58
Parthenogenesis 35
Partition 227
Peach 52, 300
Pear tree 52
Peas 51
Pests 283
Phacelia 51
Piping, queen's 41
Pistil 17
Plants, nectar 50
Play flight 11
Plum tree 50, 52, 298
Pollen
  artificial 94
  bees collecting 17
  cells containing 25
Pollen baskets of a worker bee 9
Poplar buds 18
Practicing on an empty movable-frame hive 116
Preparing supers 178
Price of
  hives 70, 96

swarms 70
Primary swarm, *see* Swarm
Priming frames 48, 100, 102
Principles applicable to all hive systems 246
Principles, general, and comparing methods 245, chapter 18 summary
Probing a hive 88
Problems with
  feeding 141
  adding the first super too early or too late 181, 182
Proboscis of a worker bee 9
Procedures
  alternative 228
  first year, fall 123
  first year, spring 77
  first year, summer 116
  overview chapter 12 summary
  second year, spring 135
  second year, summer and fall 153
  spring management of fixed-comb hives 196
  third year 160
Processing honey 202
Production of beeswax 277–280
Products 253
Prokopovich 212
Promoting beekeeping 3
Propolis 18
Prunella 50
Pupa 36
Pyment 271

Q

Quality of brood 137
Queen 21
  cage 240
  cells 28, 37
  development of 37
  drone-laying 35
  egg laying 33, 34
  finding 237
  introducing into a hive 240–242
  lifespan 32
  natural replacement/supersedure 158
  piping 41
  requeening by grafting queen cells 239
  requeening by natural swarming 238
  requeening (artificial replacement) 237
  song (piping) 41
  supersedure 158
Queenless hive 84
  fixing 236
  movable-frame 131
Quinby hive 213.1

R

Races of honey bees 242
Rape 4, 15, 50, 51, 53, 305
Regions favorable for beekeeping 49, 53
Repellents 223
Requeening
  artificial 237
  by grafting queen cells 239
  by natural swarming 238
Reseda 300
Reserves (honey) to be left for the winter 125
Returning an afterswarm to its hive of origin 113
Rietsche press 280
Robbing 92, 93, 128
Roof, vertical hive 172
Root hive 213.1
Rosemary 302
Rubus 50
Rule for priming frames with a bead of wax 102

S

Sage 50, 302
Sagot hive 212
Sainfoin 15, 51, 300

honey 257.1, 297
  nectar 297
Sale of honey 256
Salvia 50
Santonax hive 212
Savory 54
Scabiosae 303
Scales 219
Scotch hive 215
Secondary swarm, *see* Swarm
Sections 174, 191, 194
  boxes (supers) 175, 191
  frame for 194
  horizontal hives 194
  problems to avoid 193
Sections equipment 175, 191, 194
Self-heal 50
Sempervivum 302
Separators 175, 191
Settling tank for honey 129
Skep, *see* Hive
Smoker 59
  common 60
  Layens 61
  mechanical 61
Smoking 59
  fixed-comb hive 79
  frame hive 118
Soil, influence on nectar resources 55, 308
Solar wax melter 278
Solidago 50
Spacing, comb 20
Speculative feeding 231
Spring inspection, *see* Inspecting
Spur embedder 98
Stamen 17
Starting an apiary 56
Stigma 17
Stimulative feeding 231
Sting of a worker bee 9
  of the queen 21
Stings
  how to prevent 62
  preventing neighbors from being stung 63
  treating 62
Storing honey 255
Strengthening a weak hive 154
Sucrose 295
Suffocating bees 43
Sugar exudations of plants 310
Sugar paste, feeding with 232
Sulfur, treating comb with 86
Super
  adding 179–182, 184–186
  placing under a fixed-comb hive 120
  with frames for a cap hive 208
  with sections 175, 191
Supers, fixed-comb hives 45
  harvesting 188
  monitoring 183
  preparing 178
Superposition transfer 230
Supersedure (natural queen replacement) 158
Swarm(s) 39, 104, 105
  afterswarm 110
    prevention by artificial swarming 163
    collecting 111
    determining the hive of origin 112
    preventing 233
    returning to the hive of origin 113
  artificial 163
  with one hive 234
  transfer 230
  awkwardly located 105
  beginning to draw comb 40
  buying 69
  emerging 40
  emerging, different cases 114
  feeding, in case of bad weather 109
  from a swarm 114

hanging from a branch 40
installing in movable-frame hive 107
natural, how to collect 104
prevention 157
price of 70
primary and afterswarms 40
primary, determining the hive of origin 106
transporting 75
uniting 198, 199
Swarming 39
  artificial with movable-frame hives 163
  artificial with fixed-comb hives 200
  prevention 157
  season (fixed-comb hives) 197
Sweet-clover 50
Syrup for feeding 89, 90

T

Table to track hive progress 166
Tartaric acid, mead making 262
Tasks performed by bees 31, 312
Teucrium 50
Thermometer 219
Thierry-Mieg hive 212
Thistle 50, 303
Thorax of a worker bee 9
Thymes 50, 54
Toadflax 50
Toads 292
Tools for harvesting honey 129
Tracking your apiary's progress 166
Transfer
  artificial swarming 230
  direct 144
  direct, problems 152
  flipped-hive 143
  superposition 230
Transferring fixed-comb hives into movable-frame hives 142
Transporting hives 74
  swarms 75
Treatments for stings 62

Trees, nectar 52
Trichodactylus 292
Trichodes apiarius 292
Trifolium
  campestre 51
  hybridum 51
  incarnatum 50
  repens 15, 51, 300
Triungulin 292
Turnip 15

U

Ulex europaeus 295
Uncapping
  comb for extracting honey 129
  fork 222
  knife 129
  knife, two-handled 129
  rack 129
Uniting
  colonies 235
  fixed-comb hives after harvest 203, 204
  frame hives 132
  swarms 198, 199
Using
  comb 101
  honey 276
  wash water for mead 264
  wax 282

V

Value of purchased hives 66–68
Veil 57
Venom, bee 62
Ventilating bees 6
Vertical hives, different models 213.1
Vertigo 289
Vetch 16, 51, 298
Vicia sativa 16, 51, 298
Vinegar, honey 276
Violet 300
Voirnot hive 213.1

W
Warm-way hives 214
Warquin hive 212
Wasps 4, 292
Water, bees collecting 19
Water basin (trough) 73
Water-bath oil can 221
Wax 23
  allowing bees to draw, advantages of 247
  artificial/adulterated 281
  foundation 48
  foundation, advantages of 48, 119
  foundation, how to install 99
  how bees can produce more 161
  processing 277–280
  quantity of honey required to produce 161
  uses 282
Wax glands 9, 23
Wax melter, solar 278
Wax moths 290, 291
Weight
  hive 67, 68, 88, 219, 312–315
  one honey frame 124
Wells hive 213.2
Wild cherry 50
Willow 52
Wine
  ameliorated with honey 270, 271
  pomace 272
Winter chores 170
Wintering
  end of 78, 135, 160
  fixed-comb hives 76
  frame hives 134
  vertical hives 177, 190
Wood block for installing foundation into sections 192
Worker bees 7
Worker(s) 7–9
  brood 26
  cells 23
  development 36
  larvae 26
  laying 32
  number in a hive 31

Y
Young bees 31
  play flight 11

The Complete Line of Layens Equipment

We keep bees naturally in a variety of horizontal hives, and the Layens is by far our favorite model. We are happy to offer a complete line of Layens equipment for your beekeeping enjoyment.

- Layens hives (14-frame and 19-frame)
- Swarm traps (bait hives)
- Frames (fully assembled & wired)
- Bulk unassembled frames
- Jigs for assembling and wiring Layens frames
- Universal Layens extractors
- Layens frame feeders
- Entrance reducers and mouse guards
- Screen bottom boards
- Conversion kits for transferring bees from Langstroth hives
- Hive stands
- Complete swarm-catching supplies
- Plans for do-it-yourself hive and frame construction
- Books on horizontal hive management
- Educational seminars and consultations

www.HorizontalHive.com

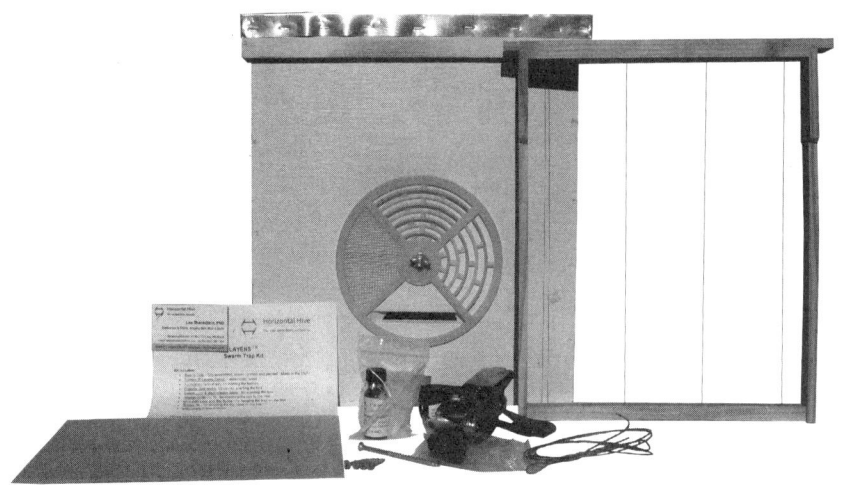

Layens Swarm Trap Kit

Start or increase your apiary by attracting free swarms of local honey bees. This complete Layens Swarm Trap Kit has all you need to succeed. It is a veritable bee magnet, used all over the U.S. Set the baited boxes on trees in the spring and see them occupied by bees! Produce honey and enhance local bee populations. This affordable model is a result of many years of successful swarm-catching.

Each kit includes:

- **Swarm trap.** Fully assembled, sealed, primed and painted. Integrated entrance gate on select models. Screened ventilation hole to prevent overheating. Interior pockets for inserting tubes of swarm lure. Top covered with long-lasting aluminum. Made in the USA!
- **Frames (6 Layens deeps).** Fully assembled and wired.
- **Foundation (wax sheet)** for priming the frames so the bees build straight comb.
- **Propolis (bee resin),** 1/2 oz, for scenting the box to make it more attractive to scout bees.
- **Swarm lure kit with slow-release tubes** for reliably baiting the box.
- **Wire & big screw** for hanging the trap.
- **Ratchet Strap (10 ft)** for attaching the box to the tree.
- **Screws (4)** for securing the top cover to the box.
- **Instructions.** Four pages of detailed step-by-step instructions based on years of experience and thorough scientific research.

www.HorizontalHive.com

Layens Hives

Our favorite horizontal hive, the classic Layens. Ideal for backyard and stationary apiaries. Fully assembled and ready to go. They come in two standard sizes:

- **14-frame hive** is recommended for southern climates (Zone 6 and up), as well as beekeepers who value bee health and sustainability over maximum honey production. It can hold 40 lb of surplus honey.

- **19-frame hive** is recommended for northern and mountainous climates as well as areas with very abundant honey flows. Can hold 80 lb of surplus honey.

Both models feature:

- **Thick warm walls** built of 1-1/2" pine—twice the insulation value of conventional hives.

- **Ventilated top** covered with long-lasting aluminum.

- **Two entrances** for making splits inside the same box and for other procedures.

- **Frames included**, fully assembled and wired. Layens frames are ideal for bees' development and wintering.

- **Optional legs** for added convenience.

- **Fully compatible** with our universal Layens extractors.

- **Comprehensive management advice** is contained in this book.

www.HorizontalHive.com

Universal Layens Extractor

The most versatile extractor ever. This is the model we ourselves use in all our apiaries. Amazing value, made to last.

- **Stainless steel throughout.** Both the tank and the rotating cage are made of high-quality durable stainless steel.
- **Truly universal design.** This is the only truly universal extractor available, a must-have for anyone seeking the freedom of keeping bees in any hive model of your choice. Accepts 3 Layens frames, 3 Deep Langstroth frames, 6 Medium frames, 6 Shallow frames, or 3 Jumbo Dadant frames!
- **One cage fits all.** You don't need to swap cages out when switching from one frame size to another.
- **Intelligently engineered** and high quality manufacture.
- **Well balanced.** We get many comments on how smooth and well-balanced this machine is. We fully agree!
- **Durable metal gears.** Unlike some other extractors which use breakable plastic gears or belts, this extractor's gears are made of durable steel.
- **You can see inside.** Beautiful transparent plexiglass cover allows you to see honey fly out of the comb.
- **Easy manual operation** requires no electricity and helps gently extract even the most delicate combs. Handle turns in a convenient position for easy operation.
- **Tangential design** for quick and complete extraction.
- **Easy to clean.** The barrel is made big enough so you can put your arm between the cage and the barrel and reach everywhere to clean it without having to take the cage out. Huge improvement and time saver.
- **Bigger barrel** also means there is greater centrifugal force so the combs are emptied more completely.
- **Legs and the honey gate valve included** with all extractors.
- **Ships fully assembled.** All you need to do is attach legs and handle using provided hardware.
- **Made in Europe** by the leading producer of Layens and universal extractors.

www.HorizontalHive.com

Natural Beekeeping Workshops
with Dr. Leo Sharashkin
Editor, *Keeping Bees in Horizontal Hives*

"Dr Leo taught a class that was totally dynamite! Amazingly awesome trouble free way to keep bees that manages them so in tune with nature." — unsolicited comment on Facebook

See dozens of Layens and other horizontal hives in action! Dr. Leo shares time-honored methods that make beekeeping simple, healthful, rewarding, fun, and accessible to all. His visually-rich presentations include step-by-step instructions, apiary visit, hive demonstrations, and more. These seminars attract participants from all over the U.S., Canada, and Europe.

About Dr. Leo

Dr. Leo Sharashkin is editor of *Keeping Bees in Horizontal Hives* and *Keeping Bees With a Smile* and a regular contributor to *American Bee Journal*, *Bee Culture*, and other major publications. He lives on a forest homestead in the Ozarks in southern Missouri, catches wild swarms and raises bees naturally in several dozen horizontal hives of different designs, including the Layens. Dr. Leo speaks internationally on sustainable beekeeping and organic growing. He holds a PhD in Forestry from the University of Missouri.

Program Highlights

- In-depth understanding of how bees live in nature.
- Keep bees naturally without interfering in their lives.
- Start an apiary for free by attracting local bee swarms.
- How to build low-maintenance bee-friendly hives.
- Horizontal hive models and their advantages.
- Complete how-tos of horizontal hive management.
- Keep colonies healthy and strong without any drugs.
- The ideal frame: how deep is deep enough?
- Help bees overwinter successfully in any climate.
- Foundation frames vs. foundationless natural comb.
- Natural swarming — boon or bane?
- Simplified beekeeping: one-box hive, no feeding, no heavy lifting, no queen excluders, no requeening.
- Produce truly natural honey for yourself and for sale.
- Unique hive products you never heard about.
- Help reverse bee decline and restore landscapes.
- "Pulling honey is all I do." One hive visit per year.

For seminar schedule and locations visit

www.HorizontalHive.com

What Others Are Saying

"This was the greatest experience I've ever had at a seminar. I feel so much more intelligent regarding bees and feel like I received a lifetime of beekeeping knowledge." — *Linn, Missouri*

"I am in shock of the value I have received! It was well worth driving over 2000 km." — *Rich, Canada*

"I feel like a born-again beekeeper." — *Paul, California*

"Fantastic crash course in keeping bees naturally. Fast-paced, action-packed, and to-the-point. You'll understand beekeeping—and how it should be done." — *Garl, Montana*

"Invaluable for all levels of beekeepers striving for the natural approach."

MORE BOOKS ON NATURAL BEEKEEPING & HORIZONTAL HIVES

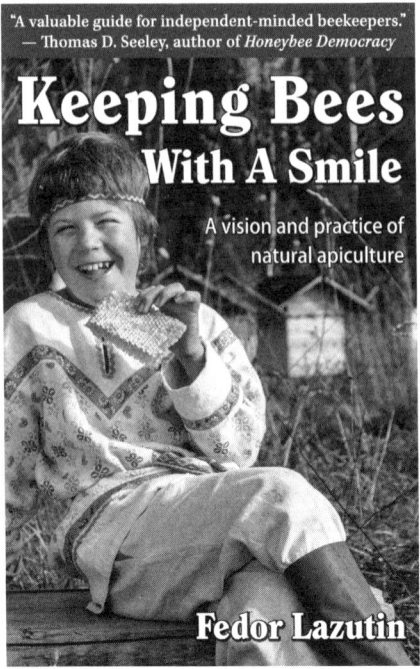

"*Keeping Bees With a Smile* is a valuable guide for independent-minded beekeepers who are seeking ways to keep bees without treating them with chemicals, disrupting their homes, and otherwise intruding on their lives. Fedor Lazutin, one of Russia's foremost natural beekeepers, describes a beekeeping system based on a trust of a bee colony as a living being capable of solving life's challenges without human assistance. Beginner-friendly and complete with fascinating photographs, it is a special book, and one that I expect will 'shake up' the thinking of the independent-minded beekeepers in North America and Europe."

— Dr. Thomas D. Seeley
professor, Cornell University
author of *Honeybee Democracy*

Are you a beginner curious about bees, or a practicing beekeeper looking for natural alternatives that work? *Keeping Bees With a Smile* is for You! Learn from an experienced horizontal hive beekeeper and discover an approach that is fun, healthful, rewarding, and accessible to all. Unique insights on how to:

- Start an apiary for free by attracting local bee swarms.
- Keep colonies healthy & strong without any drugs or gimmickry.
- Build low-maintenance horizontal hives that mimic how bees live in nature.
- Help bees overwinter successfully even in the harshest climate.
- Enhance local nectar plant resources.
- Harvest fabulous honey without stressing the bees. And *much* more.